Progress in Optical Science and Photonics

Volume 8

The purpose of the series Progress in Optical Science and Photonics is to provide a forum to disseminate the latest research findings in various areas of Optics and its applications. The intended audience are physicists, electrical and electronic engineers, applied mathematicians, biomedical engineers, and advanced graduate students.

More information about this series at http://www.springer.com/series/10091

Jayson J. Nelson

Precision Lens Molding of Glass: A Process Perspective

Springer

Jayson J. Nelson
J and T Molding Solutions, LLC
Tucson, AZ, USA

ISSN 2363-5096 ISSN 2363-510X (electronic)
Progress in Optical Science and Photonics
ISBN 978-981-15-4237-4 ISBN 978-981-15-4238-1 (eBook)
https://doi.org/10.1007/978-981-15-4238-1

This Springer imprint is published by the registered company Springer Nature Singapore Pte Ltd.
The registered company address is: 152 Beach Road, #21-01/04 Gateway East, Singapore 189721, Singapore

Acknowledgements

No one is an island, and none of us are born with knowledge. We gain information and experiences that build knowledge and wisdom in ourselves, and this often comes through the investment of others in our lives. This collection of learnings would not have been possible without the support and encouragement of others.

Many thanks to Walt Czajkowski for his hours of proofreading, edits, and suggestions. I value your opinion and insights.

I am grateful to Dr. Rongguang Liang for his encouragement to begin writing this book, and to Dr. John Pulver for sharing with me his insights and deep understanding of glass science and optics manufacturing.

And most of all my wife Tamela, for her patience, encouragement, and support, who endured long hours of endless tapping of keys on my computer. Without her support, life's adventures and accomplishments would be incomplete and empty.

I have met many people through the years, founders of companies and leaders in business and industry, and have been privileged to be mentored by some brilliant and wonderful people. But the person who has had the greatest impact on my life is my God and Savior Jesus Christ. He directs my paths in the way I should go and directs my thoughts to learn new and exciting things. To Him be all thanks, because without Him, I am nothing.

Contents

About the Author

Jayson J. Nelson has been working in optics and related industries for over 40 years. His training with precision molding technology began at Eastman Kodak Company, where he was privileged to learn from technical staff that had themselves been developing molding processes for several decades.

He has been directly involved with the development of precision lens molding capability for visible and infrared applications and has over 20 years' experience with molding technology. Current research activity is focused on controlling momentum transfer, stress relaxation techniques, surface healing, and the formation of intricate molded surfaces.

He obtained a BS in physics from Rochester Institute of Technology and is Author of eight US patents and several publications. He is on the board of OEOSC, Member of SPIE and OSA, Adjunct Research Professor at the University of Arizona College of Optical Sciences, and President of J&T Molding Solutions, LLC.

Abbreviations

AQL	Acceptable Quality Limit
CCD	Charge-Coupled Device
ChG	Chalcogenide Glass
CMOS	Complementary Metal–Oxide–Semiconductor
CNC	Computer Numerical Control
CT	Center Thickness
CTE	Coefficient of Thermal Expansion
CVD	Chemical Vapor Deposition
DLC	Diamond-Like Carbon
DOF	Degree of Freedom
FLM	Finished Lens Molding
FWIR	Far-Wave Infrared
IR	Infrared
LWIR	Long-Wave Infrared
MWIR	Mid-Wave Infrared
NIR	Near Infrared
nm	Nanometer
PID	Proportional, Integral, Derivative
PLM	Precision Lens Molding
ppm	Parts per million
PVD	Physical Vapor Deposition
REACH	Registration, Evaluation, Authorization and Restriction of Chemicals
RF	Radio Frequency Induction
RIE	Reactive Ion Etching
RoHS	Restriction of Hazardous Substances
SAG	Sagitta
SPDT	Single Point Diamond Turning
SSD	Sub-Surface Damage

SWIR	Short-Wave Infrared
T_g	Glass Transition Temperature
VC	Vitreous Carbon
VFT	Vogel, Fulcher, Tammann
λ	Wavelength of light, equal to 632.8 nm

List of Figures

List of Tables

Chapter 1
Overview of Glass Molding Processes

Precision glass molding is one of the few optical fabrication processes that is not subtractive in nature. By taking advantage of the viscoelastic nature of glass, articles can be created with little or no measurable subsurface damage.

1.1 History of Lens Molding

The process of replicating precision surfaces in glass materials has been recorded in patent literature since the early twentieth century when E. G. Johanson filed application for a "Glass Molding Apparatus" that described forming surfaces and features in glass by heating the material to a "sufficiently soft state to receive the impression" [1]. This and other early patents refer to "having invented new and useful improvements in glass molding apparatus" [2], which indicates that some technology was already in place and operational at the time of their research.

Technological advances of the past 100 years have built upon this foundation to develop what is presently known as precision lens molding. Optic manufacturing companies have invested heavily in research on materials, coatings, tooling, equipment, and production processes for advancing this technology to the state we currently enjoy.

As used today, precision lens molding is best defined as a compression molding process whereby a preshaped and predetermined volume of glass is placed between specially prepared tools to create optical quality surfaces. The components produced by these stable and repeatable processes exhibit accurate, well defined features that may be used in optical and non-optical systems.

J. J. Nelson, *Precision Lens Molding of Glass: A Process Perspective*,
Progress in Optical Science and Photonics 8,
https://doi.org/10.1007/978-981-15-4238-1_1

1.2 Principle of Operation

Precision lens molding (PLM) is a compression molding process that has application with vitreous materials. While the principles and processes presented here may be applied with plastic materials to create optical quality articles, the focus of this book will be for oxide and chalcogenide glasses. The PLM process makes use of the viscoelastic nature of glass to soften the material in a controlled manner to the point where the glass is easily deformed under the influence of an external force. Unlike other materials that demonstrate abrupt phase changes upon heating and cooling, viscous materials demonstrate a gradual and predictable transition.

When molding optical elements, the glass is heated until the viscosity approaches 10^7 Poise (10^6 Pa s). Molding is performed in a chamber under vacuum or filled with an inert gas. Oxygen needs to be removed from the system since oxygen becomes much more reactive at high temperatures (T > 450 °C) and may cause surface degradation by reducing both the glass and tooling surfaces. Once the desired temperature is reached, an external force is applied to shape the material in a controlled manner and create the desired surface. The glass is then cooled at a controlled rate to partially anneal the material and remove internal stress caused by the forming process. A typical production sequence is depicted in Fig. 1.2:

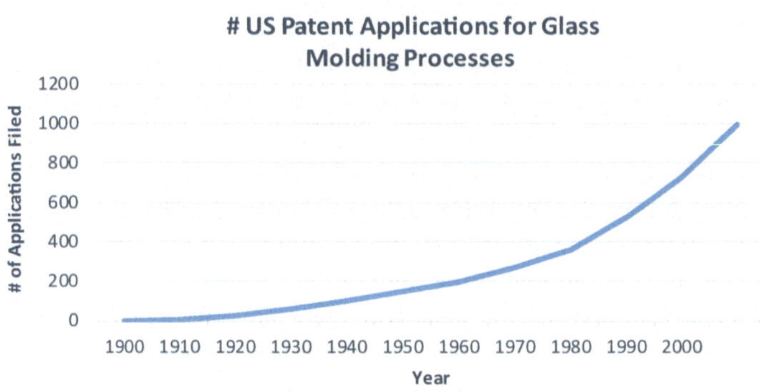

Fig. 1.1 US Patent applications for PLM processes and equipment

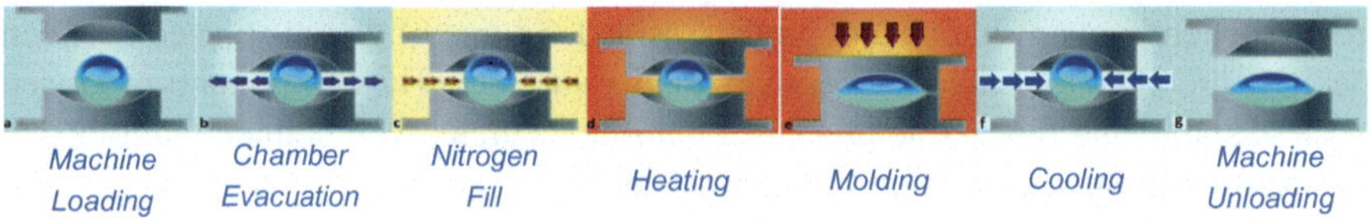

Fig. 1.2 Depiction of a typical glass molding process. *Courtesy* Fisba AG

1.3 Equipment

Equipment used to produce molded glass optics should contain the following features:

- Means to generate heat
- Means of creating and controlling a compressive force
- Precision linear motion control
- Controlled environment for molding
- Means of heat transfer to cool the system.

Prior to the late twentieth century, companies involved in glass molding were required to develop internal resources for the design and fabrication of their own custom equipment. But today, several commercial options exist for glass molding equipment that perform quite well and have made it possible for companies with less capital and internal resources to produce glass molded optics. However, while commercial equipment may be available, it remains the responsibility of the individual to develop the necessary support structure and manufacturing processes to produce high quality products. A good understanding of materials and their properties, state of the art metrology systems, thin film coatings, and glass science will produce great benefits when optimizing tool packages and developing production processes.

There are two main types of machines used for glass molding; fixed die machines and die transfer machines. Fixed die machines are those where the die package (consisting of the mold dies, mold tools, and die plates) are fixed in position throughout the molding cycle (see Fig. 2.12). The mold dies may separate to enable loading and unloading of glass articles, but the die package does not change location. One die package is used for each machine, keeping up front tooling costs to a minimum. Fixed die machines are the most common types found in industry.

Die transfer machines are those that allow the die package to be transported through different stations that correspond to the various functions of the molding cycle. For instance, a die transfer machine may have separate stations for loading, preheat, final heat, pressing, cooling, and unloading. Some specialized equipment may contain even more separation of machine utilities that isolate critical functions for better system performance, or to produce specific features in the finished optic. These machines are often used for high throughput applications where annual volumes are very high. The downside of using this equipment is that to reach maximum output and efficiency, several tooling packages must be fabricated (one for each station), which results in significantly higher start-up costs.

1.4 Process Development

Molding process development is focused on managing, controlling, manipulating, and directing thermal energy. The need for efficient thermal management with short cycle times is a main objective for every molding house, and this work begins with material selection. In general, heat transfer follows Newton's Law of Heating and Cooling that relates the rate of heat lost (or gained) in a fixed volume of material to the temperature difference between the material and its surroundings, and can be stated in equation form as:

$$\Delta T(t) = \Delta T_0 e^{-t/\tau} \tag{1.1}$$

The thermal time constant τ is a function of several material properties:

$$(\tau) = \rho C_p V / h A_S \tag{1.2}$$

ρ Density (g/m^3)
C_p Heat Capacity (J/g K)
V Volume of material (m^3)
h Heat Transfer Coefficient (W/m^2 K)
A_S Surface Area (m^2).

While Newton's Law deals mainly with convection processes, heat transfer theory based on Fourier's First Law relates to conduction processes. These equations describe the amount of heat conducted through a material per unit cross sectional area per unit time in the presence of a unit temperature gradient. The linear flow of heat in the x direction can be given by:

$$H = -KA(dT/dx) \tag{1.3}$$

H Heat Current (J/s)
K Thermal Conductivity (W/m K)
A Cross Sectional Area (m^2)
T Temperature (K).

1.5 Molding of Oxide Glasses

Many types of oxide glasses can be used in PLM processes. However, limitations imposed by commercially available equipment restrict most operations to the use of "moldable glasses". Moldable glasses are defined as those having a transformation temperature generally below 550 °C, with compositions that are compatible with molding processes, and offered in shapes and sizes that compliment molding processes. Some companies have developed equipment with greater processing capabilities and have demonstrated the ability to mold more exotic materials, but this requires a good amount of knowledge and experience, along with specialized equipment.

Oxide glasses are considered to be optical glasses that have good transmission and refractive index, with dispersion properties having application primarily in the visible portion of the electromagnetic spectrum. Some of the main concerns surrounding the glass during molding are:

- devitrification (crystallization) of the glass upon cooling
- adhesion between the glass and tooling
- formation of internal stress in the glass
- change in refractive index
- smooth flow of material during pressing.

In 2003, the European Union adopted Directive 2002/95/EC, also known as the Restriction of Hazardous Substances, or RoHS. This required changes to many glass chemistries that were previously produced with compounds such as lead oxide and arsenic pentoxide. Similar directives were eventually adopted by many of the world's leading manufacturing nations so that today a wide range of glasses are made to be "RoHS Compliant", or "Green Glasses". These advancements were extended to the family of moldable glasses so that the designer is able to make environmentally conscious choices with few restrictions.

The range of oxide glasses available for molding has grown in recent years as most of the major glass providers have expanded their offerings for these materials. This benefits both the designer and fabricator and has enabled PLM processed optics to find their way into a wide variety of optical systems.

1.6 Molding of Infrared Materials

1.6.1 Infrared Optics—Crystals

In general, optics made from infrared transmitting crystalline materials can be produced through conventional grind and polish, diamond turning and grinding, and perhaps other finishing processes that are subtractive in nature. However, PLM processes cannot be used to manufacture articles from crystalline materials since they would lose their crystal properties through the process, if they could even be processed at all.

Infrared (IR) transmitting crystals are by far the most common material choice for infrared applications. While some crystals demonstrate transmission from the ultraviolet through the visible spectrum and beyond, most applications for these materials are found in the infrared. Many have properties that are quite desirable to the optical designer, and their availability has made them almost ubiquitous for all types of infrared systems. However, the optical and physical properties of all crystals are a function of the crystal stoichiometry and structure, which may limit available options for the designer. Certain tradeoffs are often required between material properties and system requirements that may compromise overall system performance.

Transmission range is of primary concern when working with infrared materials as the infrared spectrum is much broader than the visible spectrum and transmission

Fig. 1.3 Transmission ranges of common IR materials

Material	T_{start} ***(μm)***	T_{end} ***(μm)***
CaF_2	0.13	9
Al_2O_3	0.15	5.5
SiO_2 (UV Grade)	0.18	2.5
NaCl	0.20	20
SiO_2 (IR Grade)	0.25	3.5
ZnS	0.37	14
ChG	0.50	16
ZnSe	0.50	22
CdMnTe	0.63	22
Si	1.2	10
Ge	1.8	23
Si	48	100

bands can vary greatly between materials. Other material properties such as refractive index, hardness, and thermo optic coefficient can be equally critical to a design. For the most part, optical, physical, and thermal properties of crystals are established—they are functions of the composition and a unique arrangement of atoms, whereas glass materials can be tailored for specific purposes, but with some limitations.

All materials have positive and negative attributes, and there is no single material to meet the demands of every application. For example, germanium has the highest refractive index of the infrared crystals, but also has a very high thermo optic coefficient and suffers from transmission loss at elevated temperatures. Alkali halides have broad transmission but are very soft and hygroscopic. Zinc based materials have good transmission and thermo optic coefficients but have high dispersion and scatter. These properties are inherent to the material and cannot be significantly altered (Figs. 1.3 and 1.4).

1.6.2 Molding of Infrared Materials—Glasses (Chalcogenides)

Chalcogenide materials can be readily formed into glasses, and as such, have distinct advantages over crystalline materials. The composition of these glasses can be manipulated to modify their optical, physical, and thermal properties (Figs. 1.5 and 1.6).

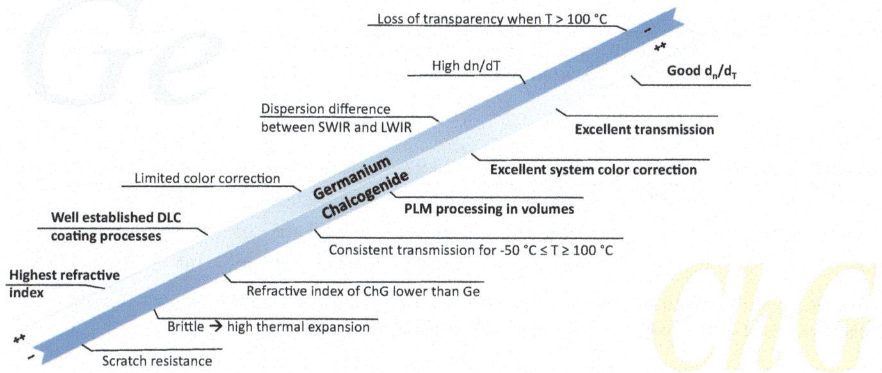

Fig. 1.4 Ge and ChG properties. *Courtesy* SCHOTT AG

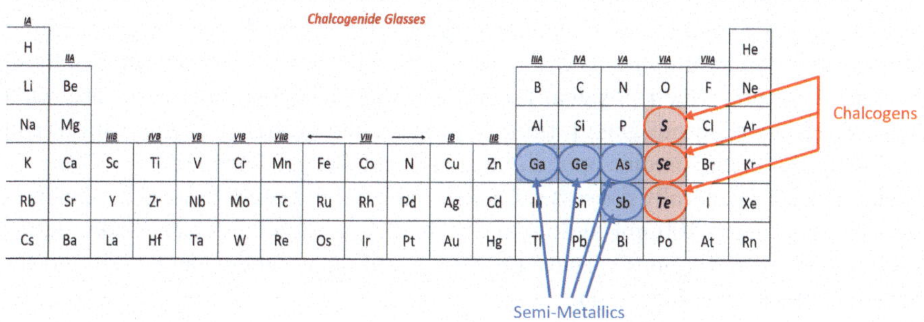

Fig. 1.5 Periodic table of the elements

Fig. 1.6 Exemplar glass forming region of ChG mixture [3]

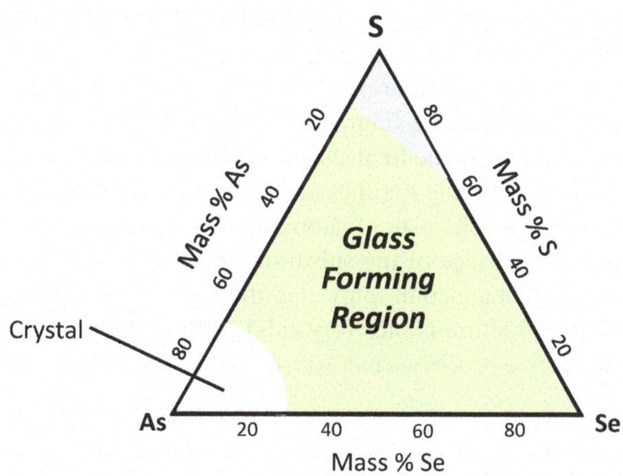

Table 1.1 Properties of some common chalcogenide glasses

	Refractive index (η)	Transmission start (μm)	Transmission end (μm)	CTE (ppm/K)	Tg (°C)	Ts (°C)	H_k (10^{-6} mm^2/N)
$Ge_{22}As_{12}Se_{55}$	2.4967	0.82	11.6	1.21E−05	368	445	141
$Ge_{30}As_{13}Se_{32}Te_{25}$	2.7869	1.28	13.9	1.34E−05	275	360	136
$Ge_{10}As_{40}Se_{50}$	2.6090	0.84	11.9	2.04E−05	225	310	112
$Ge_{28}Sb_{12}Se_{60}$	2.6032	0.96	12.1	1.40E−05	285	348	113
$As_{40}Se_{60}$	2.7782	0.90	13.2	2.08E−05	185	236	110

Refractive index measured at 10 μm
Transmission start and end approximate values, considered at 40% level

Chalcogenide glasses (ChG) have their basis in the chalcogen elements (group VI)—sulfur, selenium, and tellurium. In most cases, a chalcogen element is mixed with one or more semi metallic elements (groups III, IV, or V) such as germanium, gallium, arsenic, or antimony. The precise composition and relative amounts of these elements determines their physical properties. Since chalcogen materials are often characterized by generous glass forming regions, a large variety of compositions can be produced to create precise optical, physical, and thermal attributes.

These glasses generally possess refractive indexes between 1.8 and 3.2; transmission can range from 400 nm to 16 μm; thermal expansion can extend from below 10 ppm/K to over 20 ppm/K. These properties and many others can be tailored for specific applications (Table 1.1).

Chalcogenide glass optics can be produced by all the methods used for crystals, but because of their vitreous nature, optical surfaces can also be molded to produce finished articles. This can greatly reduce production costs, and since molding is not a subtractive process, no measurable subsurface damage is created, which results in a finer surface than one created by conventional processing.

This is a key difference between molding and conventional processing. Conventional processes are subtractive in nature—that is, they remove material from a larger volume to form the final shape. While much research has been applied to understand ductile machining regimes and reduce the damage caused by such removal, the fundamental mechanism of subtractive processing is to tear small amounts of material from the surface of the substrate. Inevitably, this causes fractures at and under the surface that are commonly classified as subsurface damage. For materials with preferential fracture planes (crystals) or low fracture toughness (chalcogenides), these subtractive processes can substantially weaken the surface.

1.7 Finished Lens Molding

Glass molding is a compression molding process that begins with a preshaped and predetermined volume of glass. No material is removed from the optical surface, and since the glass is displaced at a low viscosity condition, no fractures are created. In fact, the acceptable range of process temperatures is quite large for molding, and in some cases, a fire polished surface can be produced. This represents the strongest possible surface, approaching bulk strength values.

Finished lens molding (FLM) differs from precision lens molding in that the molded article is complete after molding and does not require additional processing such as centering or beveling. For extreme applications requiring tight control of diameter, PLM processing may be more common.

Finished lens molding is an enabling technology for certain manufacturing challenges and can also be an economical alternative to conventional processing. A study comparing production processes for 5 different meniscus lens diameters of chalcogenide materials revealed the crossover point between molding and single point diamond turning (SPDT) methods. In general, molding becomes economical with increases in production volumes and optical surface area. For larger diameter optics (20–30 mm), the crossover was less than 100 units (Fig. 1.7 and Table 1.2).

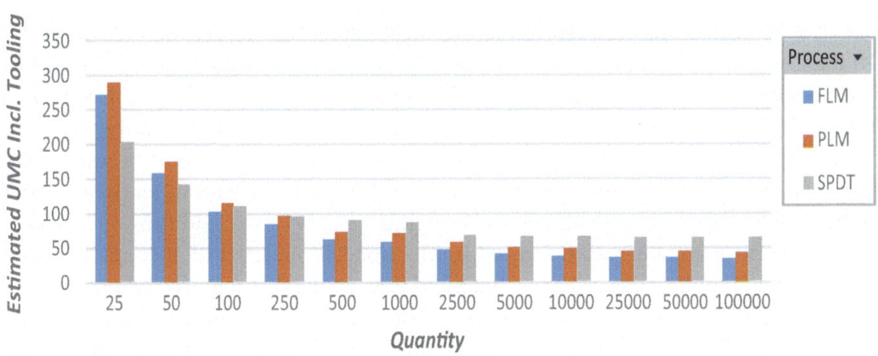

Fig. 1.7 Process cost comparison, 25 mm dia meniscus lens

Table 1.2 Crossover quantity between FLM and SPDT

Lens diameter (mm)	Crossover quantity (AU)
7.5	5000
10	1000
15	350
20	90
25	75

Approximate values shown
Assumes amortized tooling, NRE

Surface Description	Manufacturing Cost	Comments
Bi-concave; aspheric		Very high risk
Bi-concave; spherical - aspheric	High	Post processing may be required
Meniscus (spherical or aspheric)		
Bi-convex; aspheric - aspheric		Slightly more expensive than spherical - aspheric
Bi-convex; spherical - aspheric	Low	
Planar - Convex (sphere or apshere)		Rotationally symmetric

Fig. 1.8 Manufacturability guide for PLM operations

1.8 Types of Optical Surfaces

Planar, spherical, aspheric, diffractives, and arrays are all candidates for lens molding. Off axis surfaces and wedged surfaces may be molded, but with certain restrictions. Rotationally symmetric surfaces are well suited to this process. Figure 1.8 depicts surface descriptions and the potential for production by molding technology.

The lens molding process will yield best results when material flow is equal in all directions. Therefore, geometries such as rectangles, wedges, or triangles may present challenges that should cause the process engineer to perform mold flow analysis, or perhaps investigate alternate means of production. Some level of success has been demonstrated molding these shapes, but mold flow analysis is required when working with specially engineered preform shapes. Non-rotationally symmetric surfaces and the creation of fiducials are frequently discouraged for production molding processes, mostly because of difficulties encountered with mold tool manufacture, preform design, preform location, and preform orientation during processing. In the absence of adequate analysis tools, these features are often added to molded optics with standard post processing methods.

1.9 Preform Types and Manufacture

Preforms for PLM processing require a suitable surface condition prior to molding. This generally means a scratch-dig surface quality specification of 20–10 or better for visible (oxide) applications, and 40–20 or better for infrared (chalcogenide) applications. This is due to the fact that surface quality only worsens during molding as the preform surface is stretched to its new shape. Molding always results in an increase of surface area, and a smaller difference between preform shape and finished lens shape will result in a lesser impact on surface quality. Near net shape preforms may be required for certain applications.

Table 1.3 Preform shapes and general applications

Lens Geometry	Preform Type			
	Ball Preform	Plano Plano	Lenslet (Plano - Convex)	Lenslet (Bi-Convex)
Bi-Convex	•			•
Equal Meniscus	•		•	
Positive Meniscus	•		•	
Negative Meniscus	•		•	
Bi-Concave		•		

Surface roughness of the preform may affect roughness of the molded lens, but not greatly, since this metric is generally of a much lower magnitude than the physical dimensions of a scratch or dig, and low level asperities can be reduced with careful processing. Glass is not usually heated to the point of liquid flow—it is only softened so that it can be reshaped without risk of damage. In special cases and with certain materials, it is possible to flow the glass surface during pressing for complete removal of cosmetic defects. However, roughness of the tool surface will then be of concern since the glass will more closely replicate the high spatial frequency attributes of the tool.

Surface figure of the preform does not affect the final surface of the lens, except under extreme conditions. If the preform figure were to be so poor that it could impede the flow of glass during molding, then voids could be formed in the final lens surface that would be unacceptable. In general, a change in surface figure is the object of PLM processing (Table 1.3).

1.10 Process Capabilities and Tolerances

Some key physical attributes of optical surfaces are surface figure, irregularity, roughness, and quality. Mold tools are manufactured to the highest possible standard since the replicate (finished optic) can never be better than the master (mold tool). Surface figure attributes, or long spatial frequency errors, will be reproduced on the surface of the lens at process temperature, and will then contract at a rate different than the tool. This differential contraction can be used to the advantage of the molder, who can influence surface figure (power) by as much as $\pm 1.5\ \mu m$.

Surface quality, or the level of large scale scratches and digs on the surface, will only worsen with molding, and the amount of degradation depends in part on the distance the preform surface is stretched during molding. If an application calls for very tight surface quality specifications, the best path to success may be by starting with a highly polished near shape preform that limits the amount of glass movement during forming.

Table 1.4 Typical tolerances for molded chalcogenide optics

	Standard quality	Precision quality
Diameter	5–30 mm	3–30 mm
Aspheric figure error (fringes @ 633 nm)	5	2
Irregularity (fringes @ 633 nm)	2	1
Vertex radius	±1%	±0.1%
Decenter (mm)	±0.015	±0.005
Wedge (arcmin)	5	2
Center thickness tolerance (mm)	±0.030	±0.015
Diameter tolerance (mm)	±0.025	±0.010
Surface quality	60–40	20–10

Other physical attributes such as diameter, center thickness, and wedge are mostly controlled by design of the tooling package. Typical specifications and tolerances for chalcogenide molded optics are shown in Table 1.4.

1.11 Applications

Oxide glasses find application mostly as lenses and filters for medical, industrial, commercial, defense, and imaging industries. Windows, spherical lenses, cylinders, aspheres, and arrays are just some of the tools available to optical engineers. Windows and spherical optics are perhaps the most common shapes since they can easily be formed in mass quantities at low to moderate cost using traditional fabrication methods.

Aspheres, acylinders, and arrays are more expensive to manufacture than spherical surfaces since each has a distinct axis of rotation that makes mass production more difficult. Cylindrical optics lie somewhere in between. However, the advent of commercial PLM processing has enabled economical high volume production of all types of surfaces, including aspheres and arrays.

The availability of aspheric manufacturing has provided engineers with a tool to minimize spherical (monochromatic) aberrations that arise from the nonlinearities in Snell's Law while reducing the number of elements in a system. Chromatic aberrations are abated either by using glasses with proper dispersion properties or by using diffractive surfaces. Diffractive surfaces had been nearly non-existent with oxide glasses due to manufacturing difficulties, but are now becoming more common with PLM processing.

The telecommunications industry has benefitted tremendously from the development of glass molding technology. Collimator optics, micro optics, and arrays with optical surfaces less than 1 mm in diameter are all employed to accurately transmit and manipulate electromagnetic signals. Molded glass optics are nearly ubiquitous in computers and cell phones that use molded lenses in the cameras of their imaging systems.

In the medical field, devices such as endoscopes and microscopes make use of precision molded optics to collect and reproduce images.

Industries such as research, energy, and defense make use of lasers that require low surface scatter, high durability, and high resistance to both chemicals and temperature. Beam shaping, creation of flat top wavefronts, and other wavefront shaping techniques can be realized with PLM processing.

The introduction of infrared transmitting glasses over the past 30 years has opened new applications for infrared sensing and imaging. Developments in optical materials, sensor materials and fabrication, and metrology have all contributed to this increase, and many forecasters now project annual growth rates of over 20% for the next several years. New manufacturing methods such as molding are actively being sought to accommodate the demands created for quick product ramp times and shorter life cycles. These objectives have opened the way for chalcogenide materials since they offer a variety of options for physical and optical properties and are amenable to a wide range of manufacturing processes.

Infrared sensors offer high quality, passive imaging solutions without the need for an external light source. For applications above absolute zero, infrared detectors and imaging systems can provide considerable information about objects by sensing their inherent natural emissions of energy. In fact, information regarding both surface and sub-surface conditions can be obtained in this manner.

For example, military applications may focus on target recognition or detection. Equipment monitoring to detect heat buildup in circuits or tooling is a common industrial application, while commercial and industrial systems both use infrared sensing to detect unwanted heat losses in buildings. The agriculture industry uses infrared technology to detect ripeness in foods. Medical uses may focus on individual biomedical detection or medical monitoring in society.

Infrared sensing is used to see through smoke and debris in firefighting applications, and in the same manner is useful in astrophysics since IR radiation easily penetrates interstellar gas and dust clouds. The list of applications continues to expand as developments in materials and manufacturing methods enable new uses for this technology (Fig. 1.9).

Custom melts of chalcogenide materials have already demonstrated increased spectral range, higher transformation temperatures and hardness, and optimization

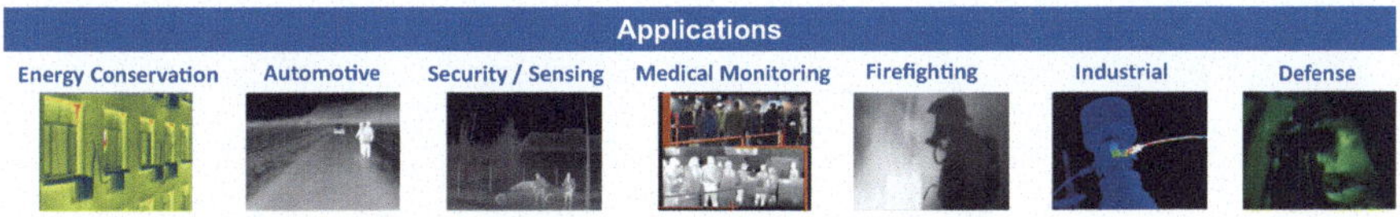

Fig. 1.9 Some current applications of infrared imaging and materials

of optical properties such as refractive index and thermo optic coefficient. This versatility is an enabling factor for increased application of precision lens molding for these materials.

1.12 Conclusion

For more than 100 years, the use of precision molding equipment with well-defined processes has enabled the production of highly complex optical surfaces for a wide variety of markets and applications. Extension of this technology to the molding of infrared transmitting chalcogenide glasses has also demonstrated good results and has facilitated imaging solutions across a broad spectral band. Continued development of materials and processing techniques promises exciting challenges and opportunities for years to come.

References

1. E.G. Johanson, US Patent 1297566, Glass Molding Apparatus, 18 March 1919
2. J.J. Wanko, US Patent 994806, Glass Molding Apparatus, 13 June 1911
3. W. Vogel, Chemistry of Glass, The American Ceramic Society, p184, 1985

Chapter 2
Tool Materials and Tooling Package Design

Choosing the proper tooling materials is fundamental to all molding. Too stringent of requirements is just as detrimental as underestimating the need for suitable material properties.

2.1 Material Requirements and Considerations for Molding Precision Surfaces

2.1.1 Traditional Tool Materials

Glass molding technology of precision optical surfaces was originally developed for the manufacture of oxide glass lenses with application in the visible range of the electromagnetic spectrum. Successful implementation of glass molding technology requires expertise with several areas of science and engineering. This section deals primarily with requirements for the materials used as molding tools and some of the more common types found in practice.

Some of the main considerations for selecting substrate materials to be used as glass molding tools are:

- Reactivity with other elements
- Machinability
- Thermal conductivity
- Thermal expansion coefficient
- Material hardness
- Fracture toughness.

Chemical, mechanical, and thermal attributes must all be considered before final selection of materials, with every material having certain strengths and every application demanding specific requirements. Thermal requirements may be the most

J. J. Nelson, *Precision Lens Molding of Glass: A Process Perspective*,
Progress in Optical Science and Photonics 8,
https://doi.org/10.1007/978-981-15-4238-1_2

Table 2.1 Thermal properties of several moldable glasses

Glass type	Manufacturer	Tg (°C)	At (°C)
L-BAL35	Ohara	527	567
L-BSL7	Ohara	498	549
L-LAH53	Ohara	574	607
L-LAL13	Ohara	534	575
L-PHL1	Ohara	347	379
L-TIM28	Ohara	504	539

demanding due to the large temperature swings encountered when molding oxide glasses (Table 2.1).

Molding temperatures for oxide glasses can vary greatly, and with the proper equipment, many of the glasses characterized by high transformation temperatures can be molded into finished optics. Most commercial equipment limits the maximum allowable processing temperature to 750 °C or less.

Few materials can withstand continued thermal cycling between 20 and 750 °C without degradation to the polished surface. Materials from the carbide family are often chosen for their ability to withstand thermal cycling, their relatively small grain size, and the capacity to retain a fine polished surface. Chemical vapor deposition (CVD) grown silicon carbide and sintered, binderless tungsten carbide are common industry choices. While these materials demonstrate strong thermal performance, they are some of the hardest materials on earth and can only be machined with diamond-based products. This limits the types of features that may be machined into the tool, and therefore, the types of features that can be replicated in the optic (Fig. 2.1).

Whereas carbides are commonly used for oxide glass molding, many metals and transition materials are suitable for chalcogenide molding due to the lower transition temperature found in most chalcogenide glasses. However, tooling materials must be carefully chosen to compliment the type of heat source employed, which will be discussed later.

Transfer of thermal energy within the system is a key consideration of mold design, and grain size of the material plays an important role in thermal transfer. Figure 2.2 depicts the thermal behavior of certain carbides based on material grain size. While larger grain size promotes higher thermal conductivity, smaller grain size provides better polishability and greater material hardness when processing conditions are

Fig. 2.1 Various grades of tungsten carbide materials [1]

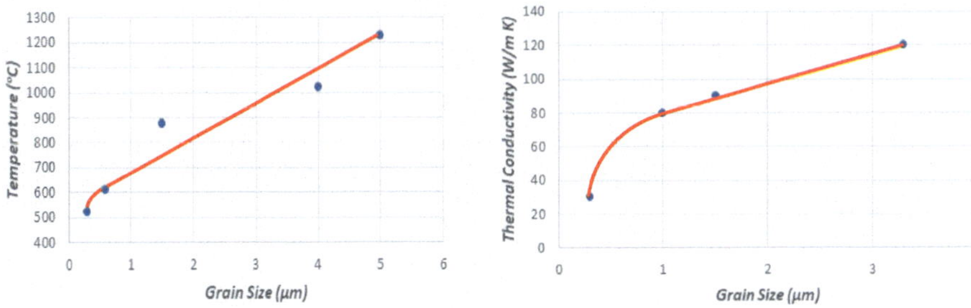

Fig. 2.2 Effect of temperature on grain size for WC compound [1]

constant. Post processing techniques can be used to densify some materials and improve their physical properties. For example, materials exposed to hot isostatic pressing (HIP) may demonstrate higher mechanical strength due to the elimination of voids between particles.

Grain size has a determinate effect on many mechanical properties of substrate materials used for glass molding. Decreasing grain size increases both the strength and the toughness of a material and is the only mechanism whereby both are increased. Other methods of increasing strength lead to a general decrease in toughness. The effect of grain size is most apparent on those properties related to early stages of deformation (ex., yield strength demonstrates a higher dependence on grain size than does tensile strength).

In regard to producing optical quality surfaces on the tool, grain size becomes a primary consideration. Smaller grains lead to better polishability and lower surface roughness. Over time, thermal fatigue from repeated cycling will tend to increase grain size by promoting rapid coarsening of the grains in critically stressed regions, so materials that are resistant to this effect will demonstrate the longest tool life and lowest overall costs.

Material hardness can be considered a key attribute of mold tools since it is directly related to scratch resistance and surface durability. However, most tools will have protective coatings applied to the optical surface so that substrate hardness becomes less of an issue and the focus shifts to properties of the protective coating.

2.1.2 Thermal Properties of Tool Materials

Table 2.2 lists thermal properties of materials commonly used in oxide glass molding operations. It is easily seen that for applications demanding very low thermal expansion properties, amorphous carbon may be a good choice. However, the poor thermal conductivity of a-carbon may result in long cycle times and low throughput, which can diminish its attractiveness as a tool material. CVD grown silicon carbide has superior thermal conductivity properties, but difficulties and limitations with post processes such as grinding and polishing should be considered. Certain metal alloys

Table 2.2 Common materials used in oxide glass molding operations

	Density (ρ)	Cp (J/g K)	K (W/m K)	κ (m^2/s)	CTE (ppm/K)
WC	15.40	0.20	82.0	2.66E−02	4.90E−06
CVD SiC	3.21	0.66	250.0	1.18E−01	4.50E−06
a-Carbon	1.51	0.70	5.8	5.49E−03	2.10E−06
Graphite	1.84	0.71	84.0	6.43E−02	6.00E−06

are easy to machine but may have less desirable thermal properties. Efficient designs will result from careful analysis of all aspects of the particular application and the proper use of different materials within the tool package.

Volumetric thermal diffusivity κ (m^2/s), is a measure of thermal stability for non-homogeneous temperature distributions within the molding tools. Materials with high diffusivity values are preferred since they transfer heat more rapidly, which allows them to quickly achieve equilibrium with their surroundings. It is also suggested that increased thermal transfer between the mold tool and the glass may result in less distortion of the molded article, which is attributed to the viscous nature of glass and the introduction of thermal stress in the glass caused by temperature gradients during cooling.

Thermally induced stress in the glass σ_{th}, can be calculated as [2];

$$\sigma_{th} = f \cdot \frac{(\alpha \cdot E)}{(1 - \mu)} \cdot \Delta T \quad (2.1)$$

f Glass specific factor
μ Poisson's ratio
α Thermal expansion coefficient
ΔT Temperature difference
E Young's modulus.

It is shown that σ_{th} varies linearly with the thermal gradient, so minimizing ΔT across the tool surface results in a corresponding reduction in stress within the optic, which makes process development much simpler. The other factors in this equation are material dependent and combine to what is known as the thermal stress factor of the glass. The thermal stress factor ϕ_w (MPa/K), is specific to the glass composition and is defined as;

$$\varphi_w = \frac{(\alpha \cdot E)}{(1 - \mu)} \quad (2.2)$$

Glasses with a high thermal stress factor will require increased process development since they are more susceptible to deformation upon cooling and may be more fragile during post processing operations such as centering or coating. Once stresses are formed within the glass article, they can only be removed by post annealing. The danger with this approach is that the surface of the optic may change as stress is released, often resulting in surfaces that are no longer within specification. Therefore, it is always best to design the molding process such that internal stress in the glass is minimized.

Other detrimental effects of stress within an optic are changes in refractive index as described by photoelastic coefficients, and stress birefringence as defined by the stress optical coefficient. Accurate thermal modeling and careful thermal management across the entire tool package are valuable tools that yield positive results in the finished optic.

2.1.3 Compatibility with Heat Source

Tooling materials for both mold dies and their support structures must be matched for the type of heat source employed. The three main types of heat generators used for molding applications are; infrared lamps, RF induction, and resistive heaters. Each of these systems has benefits and drawbacks, and best system performance will be achieved through careful selection of tooling materials.

The fundamental processes of heat generation for these systems are; radiation and absorption (infrared lamps), electron excitation within the tool package (RF induction), and electron excitation outside of the tool package (resistive heat source). Once thermal energy is created, the primary source of transfer is conduction, although convection provides some contribution due to heat energy radiating from the tooling surfaces to the surrounding atmosphere, which in turn serves to heat both tool and preform surfaces (Figs. 2.3, 2.4 and 2.5).

Fig. 2.3 Infrared lamp heaters. *Courtesy* Fisba AG

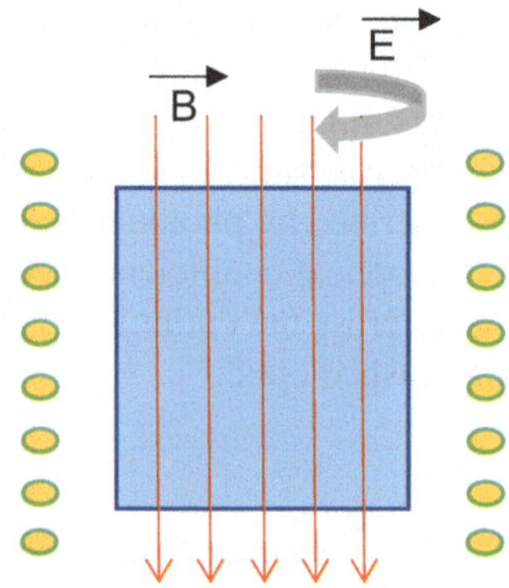

Fig. 2.4 RF Coil heaters, induction process

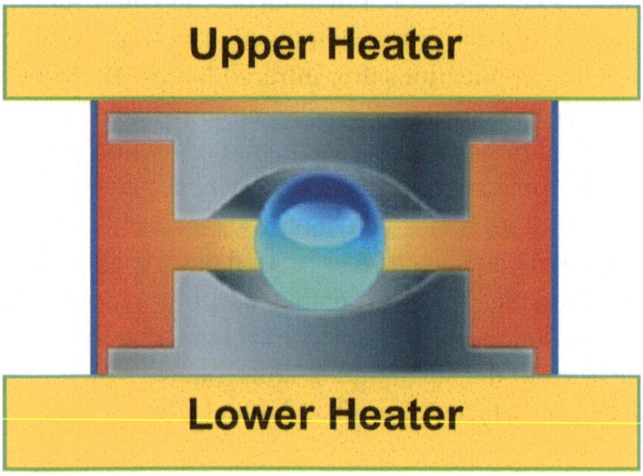

Fig. 2.5 Resistive heaters, conduction process. *Courtesy* Fisba AG

2.1.3.1 Infrared Lamps

This type of heat source uses infrared radiation to heat the tool package. The heat source does not make direct contact with the tool package and resides outside of the controlled atmosphere of the molding chamber. Since the lamps function by radiating energy from the lamps, through a quartz tube, and to the mold tool package, tooling materials must be able to absorb radiation at the wavelengths emitted by the lamps. The quartz tube specified by the equipment manufacturer is chosen to transmit the maximum amount of electromagnetic energy at the wavelengths output by the source. Figure 2.6 shows data based on 100% power output for one manufacturer of infrared lamps and displays peak power output at approximately 1.4 μm. Lamp temperature for this unit is 2200 K. Based on this graph, the most effective tool package materials will demonstrate good absorption of radiation from 0.8 through 2.3 μm.

Since materials such as chalcogenides are designed to transmit infrared radiation, little if any heat is generated directly within the glass from the lamps. Instead, the primary source of heat to the glass will be conduction through the support members and mold dies, and by convection (if non-vacuum molding processes are used). Most

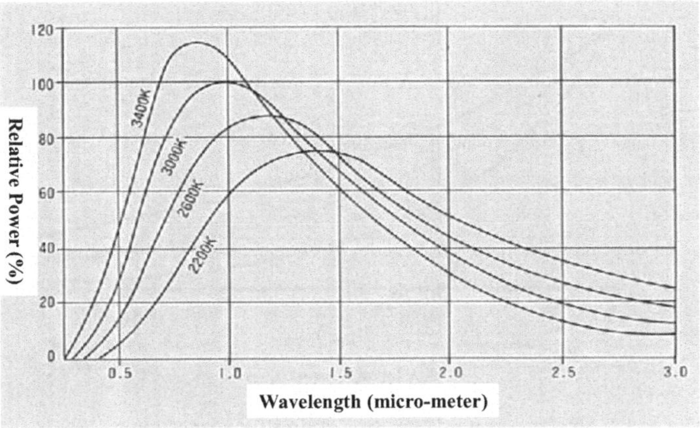

Fig. 2.6 Spectral power output of IR lamps

oxide glasses transmit poorly for the infrared, and some direct heating of oxide glasses may occur. However, the main source of heat transfer will still be conduction for both oxide and chalcogenide glasses.

Generation of thermal energy by infrared radiation occurs primarily by absorption of incident radiation and subsequent conduction through the support tooling, molds, and glass preform. This is not to minimize the effect of convection, and cycle times will always be less when the heat cycle is performed in a non-vacuum, inert environment. Vacuum molding is often a beneficial step for producing precision articles, but maintaining vacuum throughout the heat generation process is often unnecessary and inefficient.

Consider a tungsten carbide tool package that is 100 mm in diameter and 15 mm thick for both upper and lower tools. Assume that thermal ramp occurs in a nitrogen filled environment, and the separation between upper and lower tooling is small enough so that absorption only occurs on the circumference of the mold die (see Fig. 2.12).

Graphic representation of convection heating for the tool package described above is shown in Fig. 2.7 for some common mold materials,

We see that in time, all curves tend to converge to a minimum value. However, even though the initial condition is the same for all materials, the final condition remains significantly different within a reasonable amount of time.

Consider a process with a molding temperature of 520 °C that begins at a room temperature of 20 °C and uses convection as the primary means of thermal transfer. At some point early in the process, the surface nearest the heat source will experience an increase in temperature while surfaces further removed will be nearer room temperature. Temperatures in the surrounding environment will increase based on heat radiating from the warmed surface, which will transfer heat through convection to the surfaces further from the original heat source.

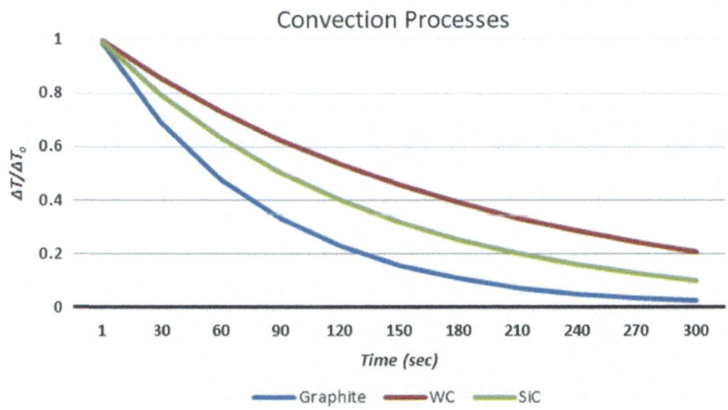

Fig. 2.7 Change in thermal gradient as a function of time for convection processes

Considering convection as the only process for thermal transfer, the temperature gradient between inner and outer surfaces with tungsten carbide tooling will still be 100 °C after 300 s. Clearly, additional thermal transfer processes must be present to enable efficient molding, but this thought experiment provides some insight into the contribution of each thermal transfer mechanism.

What makes carbon graphite perform so much better than tungsten carbide in this example? Since tooling dimensions were identical for all materials, physical properties of the materials must take precedent (a detailed discussion of fundamental equations is provided in Chap. 3). For convection, the product of density and specific heat is a factor for relative convection efficiency, which produces a much smaller value for carbon graphite than for tungsten carbide, and based on Eq. 1.1, promotes better heat transfer.

Heat transfer through conduction follows Fourier's Law, which defines the flow of heat energy relative to certain physical and geometric properties of the system. The rate of heat conduction through a medium may vary with both position and time, and therefore calculations can become quite involved. For conduction, the relevant physical property is the thermal diffusivity of the material, which itself is a function of density, thermal conductivity, and heat capacity. The figure below provides an example of calculations for temperature variations across a 10 mm diameter IRG-26 ball preform when convection and conduction processes are combined to heat the tungsten carbide tooling package shown in Fig. 2.8.

These graphs show all locations converging to a common temperature within a reasonable amount of time. The vertical distance between curves is the thermal gradient found in the preform at any point in time and should be minimized for production of accurate surfaces.

This example assumes an IR lamp source for heat generation with both convection and conduction processes unrestricted for thermal transfer. However in real systems, the physical properties of most materials will change with temperature, which results in a difference between actual data and the calculated values shown above. A safety

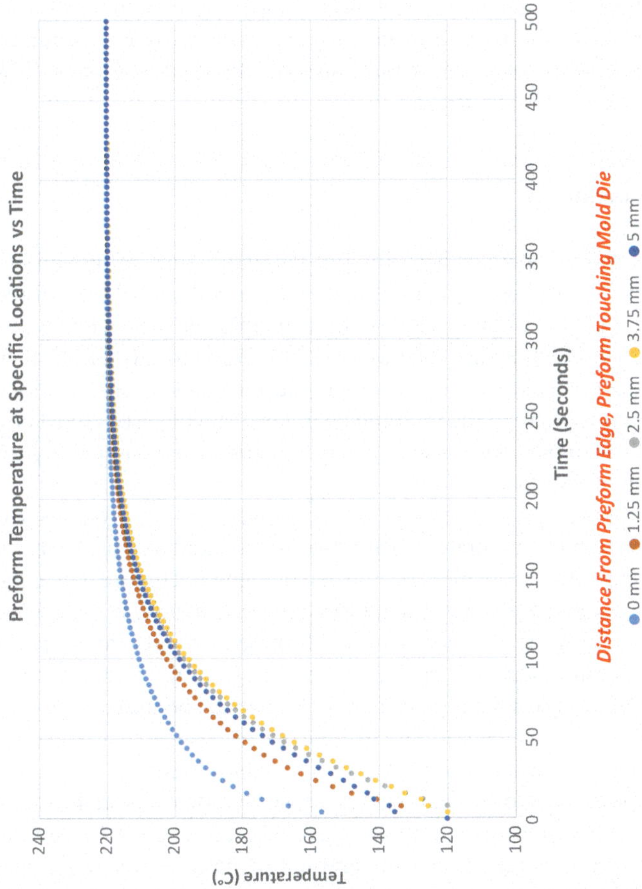

Fig. 2.8 Preform temperature as a function of time and distance for 10 mm diameter preform

factor should be added to calculated soak times to ensure that thermal gradients are at a minimum.

Another assumption is that the absorption of energy from the IR lamps and the conversion of this energy to heat is always at a maximum. Conduction will regulate the rate of energy transfer through the medium, but we assume that the amount of heat input is never less than what can be transferred. Many molding machines are equipped with PID controllers that regulate lamp output to both minimize their maximum operating cycle and ensure that temperatures do not significantly exceed their set point. This feature may result in longer cycle times than those shown above.

An understanding of the type(s) of heating experienced by each component in the mold package and the parameters that influence thermal transfer will help the designer when selecting tool materials.

2.1.3.2 RF Induction

This type of heat source operates according to Faraday's Law of Induction, which states that a temporal varying magnetic flux creates an induced electric force when operating within a closed circuit. This is the principle on which many solenoids operate for electrical components such as switches. As with infrared heat sources, the heat source does not make direct contact with the tool package and resides outside of the controlled atmosphere of the molding chamber. Tool materials must be chosen with adequate values of magnetic susceptibility and electrical conductivity to enable efficient operation of the system.

The use of ferromagnetic materials should be avoided in these systems, except in special cases. If ferromagnetic materials are used in the mold tool package, they will concentrate the magnetic flux while the non-ferromagnetic components experience very little magnetic activity, resulting in uneven heat distribution in the tool package and difficulty controlling the glass temperature. If these materials are used in any part of the support structure, localized "hot spots" will form in the tool package, which places serious limitations on the type of materials that can be used.

A positive attribute of RF generation is that the electrical current produced through induction is formed within the tool package. All other systems generate heat from without the tool package and thermal energy must be either radiated or conducted through the various components and to the glass. Therefore, an RF system is much more efficient at heat generation, and the amount of input power required will be much lower—in some cases as much as 65% lower. Since heat is generated within the tool package, cycle times will also be lower, leading to lower overall costs of production. As with all systems, tool materials must be matched to the heat source for efficient and effective operation (Table 2.3).

In one instance, a multi cavity mold with 25 mm diameter mold tools was being used in an RF heating system and a glass that had a short working range (see Sect. 3.2.3). Difficulty was encountered producing good surface figure, and when the tool package was examined, it was determined that a 40 °C temperature difference was being created across the preform surface; the outer temperature of the glass

Table 2.3 Properties of some common tool materials used for glass molding

	Density (ρ)	Cp (J/g K)	K (W/m K)	κ (m^2/s)	CTE (ppm/K)	Xm	ρ (Ω − m)
WC	15.40	0.20	82.0	2.66E−02	4.90E−06	1.00E−05	2.00E−08
CVD SiC	3.21	0.66	250.0	1.18E−01	4.50E−06	−1.28E−05	1.00E+00
a-Carbon	1.51	0.70	5.8	5.49E−03	2.10E−06	1.00E−04	3.50E−05
Graphite	1.84	0.71	84.0	6.43E−02	6.00E−06	−6.00E−06	1.35E+01

was 40 °C warmer than the glass temperature that was only 25 mm further inboard. The large temperature difference was caused by the relatively low penetration depth of current being generated within the upper and lower mold dies in the RF system. Penetration depth is inversely related to operating frequency (penetration depth $\propto$ $1/\text{frequency}^{1/2}$), which for RF systems is typically in the range of several kilohertz.

This large temperature difference caused the glass viscosity to vary across the preform surface so that the outer portion of the lens was easily formed while the inner portion remained above the preferred molding viscosity. This resulted in poor surface profile across the lens surface that produced optics that were out of specification. The solution came by interrupting the current flow around the circumference of the mold dies with carefully sized slits, thereby forcing the current to move nearer the inner regions of the mold cavities. Once completed, the measured gradient on the mold dies dropped from 40 to 5 °C. Maintaining consistent glass viscosity is a common problem when substantial thermal gradients are encountered during molding (Figs. 2.9 and 2.10).

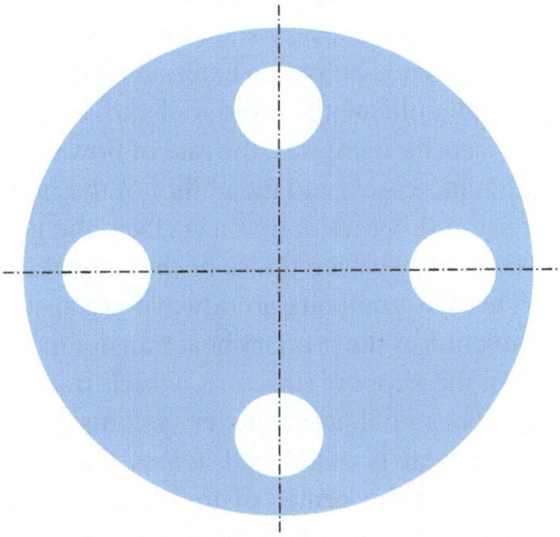

Fig. 2.9 Typical 4 cavity mold die

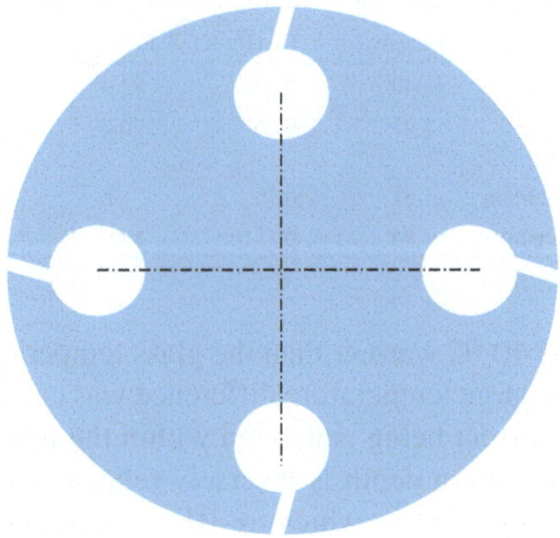

Fig. 2.10 Mold die with slits

2.1.3.3 Resistive Heaters

These are generally cartridge or plate type heaters that are normally mounted above and below the mold tool package. The tool package located within the molding chamber is in direct contact with the heat source. Heat is conducted through the support structures and into the glass, much like the operation of a hot plate or electric stove. The advantage of these systems is that they are relatively easy to control, and they can minimize unwanted thermal gradients in the glass.

Managing thermal gradients is important due to the viscous nature of glass. Viscosity is a function of temperature, and precision molding depends on accurate control of glass viscosity at every stage of the process. In the precision molding process, glass is first softened before an external force is applied to press glass against the mold surface and replicate the mold tool surface onto the optic. But as temperature decreases and viscosity increases, the rate of flow decreases, the resistance of the glass to deformation increases, and the ability of the ram to shape the glass is impacted. When the temperature varies across a surface, the glass may not uniformly replicate the mold surface, and surface figure of the optical component will be difficult to control.

Thermal gradients are always greatest in the direction of the heat source. With conduction as the primary heat transfer mechanism, thermal gradients are first formed within the support structure, which transfers heat to the mold tool and then to the glass. For radial heat transfer systems such as infrared lamps and induction sources, this gradient is evidenced across the optical surface of the tool from the outside diameter to the center of the optic. For axial heat transfer systems, this gradient occurs along the optical axis of the molded lens that lies between the two molding surfaces. Axial heat transfer systems are preferred when molding wafers, high aspect ratio optics, or articles with extreme specifications for control of optical power and irregularity.

Most optics have an aspect ratio much greater than unity. Resistive heat systems are preferred for high aspect ratio optics since thermal gradients are formed along the short dimension of the lens. Because the gradient magnitude is a function of length, the result is an overall thermal difference across the optical surface that is much less than with radial heat transfer systems. Machines with this type of heat generation find application with wafer level manufacturing and high throughput die transfer systems.

Radial thermal zones can be controlled either directly in the hot plate or by careful spacing of cartridges, and each zone can easily be regulated by the thoughtful location of thermocouples. In this manner, precise temperature feedback and control can be achieved in two directions within the tooling package.

2.2 Tool Package Design

As defined here, the tool package consists of the mold tool, mold die, and die plate. The mold tool contains the optical surface to be replicated; the mold die houses the mold tool(s); the die plate backs up the mold die and serves to house the thermocouples and mount the tooling package to the upper and lower portions of the pressing system.

The main function of the mold tool is to provide a high quality optical surface that may be reproduced by the glass. It is expected that these tools will maintain this high-quality surface for a significant number of cycles. To protect the surface from damage and wear, protective coatings are often applied that also act as a release agent for the molded glass article (see Chap. 4). The best tools will be easily machined, capable of possessing a high-quality surface, and able to withstand thermal cycling without degradation of the optical surface or change to their inherent mechanical and thermal properties.

The mold die serves to accurately locate the upper and lower mold tool(s) and to provide a mounting structure for the die plate. It is also used to locate thermocouples near to the glass during molding. This is important since closer mounting provides more accurate temperature data, which enables precise process control. The mold dies also contain locating pins that ensure proper alignment between upper and lower tools upon molding. Without these locating pins the molds could become misaligned and damage the tooling.

The die plate (*not shown in* Figs. 2.11 and 2.12) is the interface between the

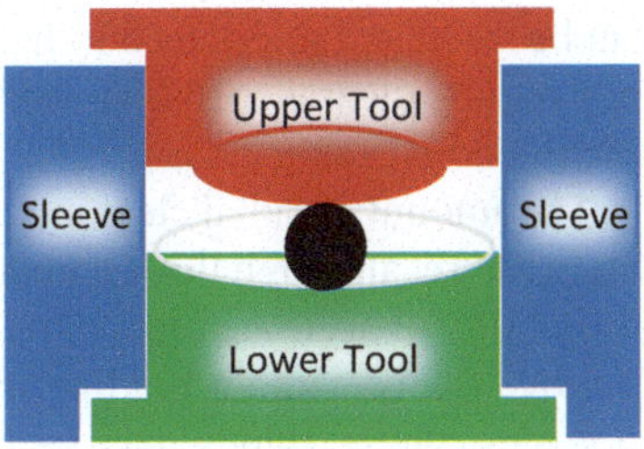

Fig. 2.11 Typical single cavity mold assembly

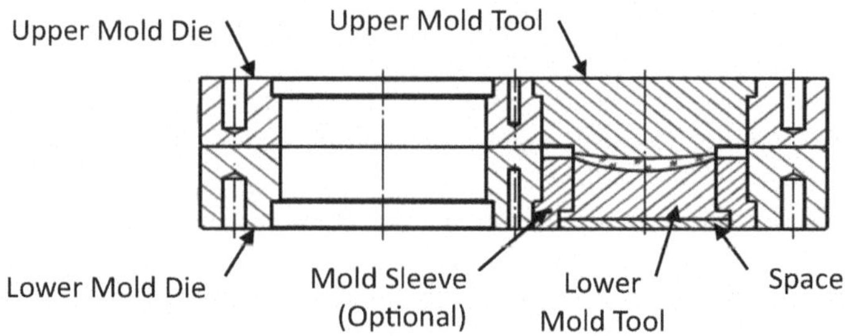

Fig. 2.12 Typical multi cavity mold tool set (only one cavity populated)

mold die and the machine. It also serves to back up the mold tool(s) in the mold die. Without the die plate, the mold tools could easily separate from the package and be damaged. Once assembled, these components create a tool package that can be safely transported and mounted in the machine.

The simplest design of tooling package is the single cavity mold. Here a common bore is created in a sleeve to house both upper and lower mold tools. The lower tool is assembled with the sleeve, the preform is introduced between the lower and upper mold tools, and the upper tool is inserted. The entire package is placed on the die plate in the molding machine where the glass is heated and pressed to form a finished optic. These designs are useful for low volume applications, have lower initial tooling costs, and are typically associated with shorter cycle times than multi cavity systems since they possess a lower thermal mass.

Multi Cavity Systems lend themselves to automation and are capable of producing a greater number of parts per cycle, but also have longer cycle times due to an increased thermal mass. Figure 2.12 depicts the cross section view of a multi cavity mold die. Lower unit process times and costs may be realized with multi cavity molds, however, the benefit is not one to one. A two-cavity system will not cut costs in half since process times will be longer and tooling costs greater. A general rule of thumb is that a doubling of cavitation by changing from a single cavity to multi cavity system will result in a 33% drop in overall process costs. Once the maximum diameter of tool package is in place, subsequent increases in cavitation will produce a benefit near unity.

Some or all of the radial clearance that is required for room temperature assembly can be eliminated at molding temperatures. Every material has different thermal expansion characteristics, and once the process temperature is known, components can be sized and toleranced to increase or decrease radial clearance. While one tool may require clearance to float within the mold die cavity, the other tool does not need this freedom (Table 2.4).

The amount of radial clearance applied is typically between 1 and 3 μm at molding temperature, although this number may vary depending on the materials used and the availability of accurate machining. This amount of clearance will allow the mold tool to move freely without binding in the mold die. However, the tolerances necessary to achieve these values may be too tight for some materials or certain machining

Table 2.4 Sample tool package design calculations

Step 1: Core diameter	
Enter Desired Lens Diameter	3.5
ChG Lens Diameter at Tp	3.515
Lower Core Diameter at Room	3.504
Clearance for Upper Tool	0.004
Upper Core Diameter at Room	3.500
Step 2: Sleeve bore diameter	
Lower Tool Clearance at Tp	0.000
Sleeve Bore Diameter at Room	3.509
Step 3: Outer sleeve diameter	
Sleeve Clearance at Tp	0.001
Enter Max Die Diameter	36.001
Die Diameter at Tp	36.053
Outer Sleeve Diameter	35.984

Step 4: Tool Heights	
Enter Max Upper Die Height	15.000
Enter Max Lower Die Height	18.000
Max Upper Die at Tp	15.023
Max Lower Die at Tp	18.026
Enter CT of the lens	1.1
Enter Total Height of the lens	1.1
Enter Desired Parting Drop	0.1
Lower Core CT at Tp	16.821
Enter Parting Line Spacing	0.5
Upper Core CT at Tp	15.628
Calculated Core Heigł	33.549
Calculated Tool Height at Tp	33.549
Lower Core CT at Room	16.761
Upper Core CT at Room	15.575

processes. If the mold tools or mold die cavities are not machined round, or the materials are too soft, then additional clearance may be required.

Radial clearance between the mold tool and mold die may be evidenced in the molded lens as concentricity error (decenter) between the optical surfaces, so care should be exercised when designing clearance into the tool set. Optical wedge may also be present and can be minimized by manufacturing tolerances and tool package design. Additional sources of manufacturing errors may be runout between body diameter and optical axis of the tool due to location tolerances between upper and lower mold dies. Each of these must be controlled and understood at room temperature and process temperatures, with tolerances based on the final specifications of the molded optic.

The simplest way to minimize wedge in the finished lens (assuming the mold tools are accurately made) is to increase the length of both the mold tool and mold dies. The wedge angle is simply a function of the tangent between mold tool and die (clearance between tool and die = opposite leg, tool length = adjacent leg) (Fig. 2.13).

While it may seem that this should be common practice in mold design, the penalty comes in process cycle time. The longer tool package contains a higher thermal mass and therefore requires more time to heat and cool. Since most molding operations are concerned with maximizing throughput, longer tooling is generally reserved for applications that require tight wedge specifications.

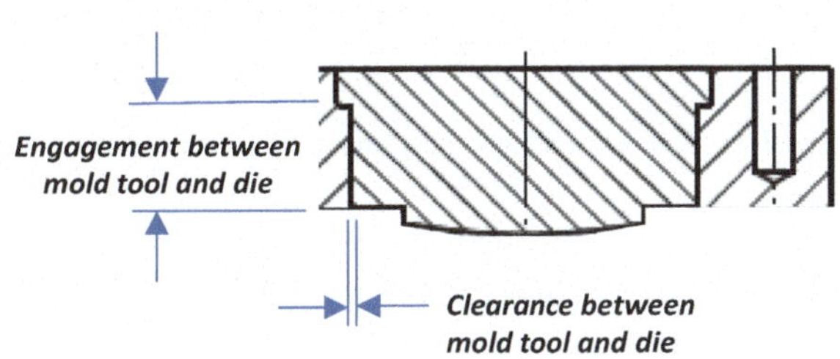

Fig. 2.13 Mold tool and mold die

2.2.1 *Tool Design*

Two main theories exist as to the best practice of tool design, and since either method can produce accurate molded optics, the choice should be based on application and requirements of the finished optic.

2.2.1.1 Closed mold

Closed mold tool design is where the two halves of the mold tool package press tightly against each other during the pressing cycle and determine the center thickness (CT) of the molded lens. Each component in a multicavity tool package has some variability in thickness due to manufacturing tolerances, which constrains the interchangeability of mold dies into different cavities.

Companies that perform glass molding will often speak of the "molding plane", which is an imaginary plane defined by a common feature (parting line of the mold dies, lower vertex of finished lenses, or some other feature). The molding plane in each cavity may vary relative to a fixed datum (usually the mold plate—mold die interface), so spacers are used behind either upper or lower mold dies for each cavity to establish a common molding plane and obtain the correct CT of the finished optic. This may require several iterations to obtain the proper value for each cavity.

In closed mold processing, the pressing ram is under force control and a predetermined maximum pressing force is applied to the glass. The upper and lower halves continue to move towards each other until they come into contact and form the parting line of the mold tool assembly. In this design, the mold tool may lose contact with the glass if the glass contracts faster than the mold package during the cooling phase. This should be seriously considered when developing the manufacturing process (Fig. 2.14).

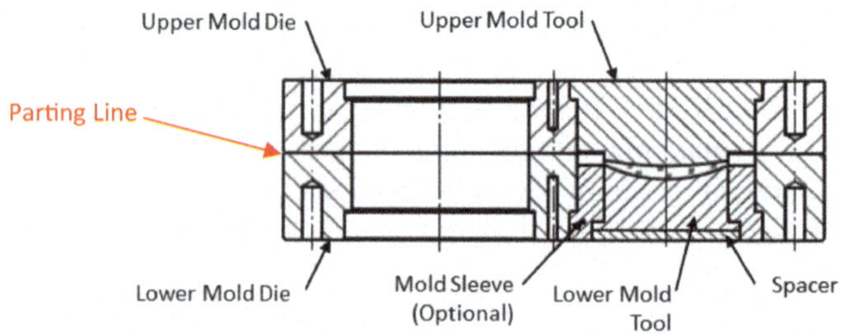

Fig. 2.14 Closed mold assembly and parting line defined

2.2.1.2 Open mold

Open mold tool design is where the two halves of the mold tool package never meet during the pressing cycle, and center thickness of the molded article is determined by the position of the pressing ram. The parting line is never defined in this type of design, and precision creation of the final center thickness relies on the repeatability of the molding equipment. While most higher quality machines will control position within a few microns, maintaining consistent temperature set points in the system will play a factor in the overall dimensions formed in the optic.

A common goal for glass molding practitioners is that the molding plane be as singular as possible; that is, that each of the defined features in a multi cavity system lie on the same plane. By creating and controlling the molding plane, we can accurately predict that the CT of the finished optics will conform to some value and tolerance. In open mold processing, manufacturing tolerances of the mold dies are the key variables for controlling CT of the molded lens. Variation among the finished articles will reflect individual variations in machining tolerances of the mold dies. Oftentimes, tool center thickness is controlled to an even finer degree for open mold processing than for closed molding.

Single cavity molding (sometimes referred to as "common bore molding") may be performed successfully and efficiently with either open or closed mold processes. Interchangeability of tools based on proper CT is not a concern for single cavity molding.

2.2.2 To Shim or Not to Shim?

Many factors will influence the decision whether to use spacers behind the mold dies for defining a common molding plane. The number of cavities, tolerance on the finished optic, machining accuracy of available tooling, and the positional accuracy and precision of molding equipment are just some of the factors that contribute to the decision. In the end, both designs are valid, and the decision should be based on the designer's confidence in the strategy they choose.

When producing high volumes of molded articles, process control is critical so that acceptable quality limits (AQL) and statistics can be used to qualify conformance. In many high volume applications, spacers are used behind the lower mold die to simplify set up for when a single mold die in a multi element package needs replacement.

In the end, it comes down to where the engineer decides to place the tightest tolerances. When spacers are used, center thickness of the mold die is not critical since the final height will be set by the spacers. In cases of single cavity molding, the outer sleeve may be used as a positive stop to set center thickness of the molded optic, or the machine itself may be used when position control is employed in place of force controlled molding.

Fig. 2.15 Chalcogenide meniscus lens having free diameter molding features

2.2.3 Free Diameter Molding

Free diameter molding is sometimes referred to as Precision Lens Molding (PLM), and articles formed by this method are identified by smooth, rounded edges and an overall look of fire polished glass. This process is used when post processing is required to create features in the molded lens that are not easily formed by molding, or when the diameter of the molded lens is not critically sized. Tight control of preform volume is not essential and typically results in a lower cost preform, but post processing may be required that drives up the overall cost of manufacture.

Tolerance on the molded diameter is not tightly held (often to around $\pm$ 0.1 mm), as final dimensions and specifications are created with ancillary processes. Notice the rounded edge on the diameter and overall look of fire polished glass on Fig. 2.15.

2.2.4 Mold to Diameter

Molding to a finished diameter is sometimes referred to as Finished Lens Molding (FLM), and articles produced by this method have crisp, well defined features. As the name implies, these components need no further processing to meet specification. Maintaining accurate preform volume is critical, as tolerances of the finished article determine tolerances for the preform. Preforms used for this method are generally more expensive, but no post processing is required, which may reduce overall costs of manufacture.

Diameters of both preform and molded lens are held tightly, and sharp edges are formed where the glass presses against the inner diameter of the mold cavity. Safety bevels are not usually created, and diameters are typically held within $\pm$ 0.025 mm. This process works well when a small volume of glass is being displaced but can also be effective for larger displacements of material. Notice the crisp, sharp edge to the diameter in Fig. 2.16.

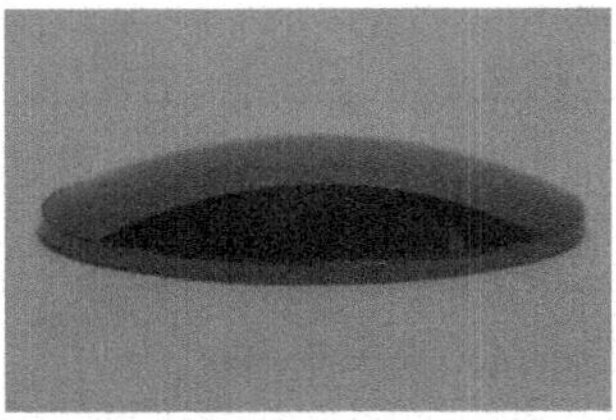

Fig. 2.16 Chalcogenide meniscus lens having mold to diameter features

2.3 Fixed Die and Die Transfer Systems

So far, we have only considered fixed die molding systems—systems where the tool package remains stationary throughout the process. Part loading and unloading both occur at the same location in the machine. Another method used for high throughput molding is the die transfer system whereby the tool package is transported between stations to reduce the unit cycle time.

These systems usually have a load station, multiple heat stations, press station(s), several cooling stations, and an unload station. By separating the process into segments, the unit cycle time equals the longest segment time in the process. Since heating and cooling normally account for the greatest amount of process time, these processes are often broken down into even smaller portions. These machines are quite versatile as both single cavity and multi cavity tool packages can be used.

Examples of these systems are the Toshiba Model GMP-310V-3R and GMP-1013-10S. The 3R system has separate zones for heating, pressing, and cooling, while the 10S system has 10 distinct zones with 4 separate heating zones and 3 pressing zones. Each zone is equipped with a separate control system over parameters relevant to that operation.

The benefit to this type of system is the high throughput and potential for lower unit process costs. The penalty is that multiple sets of tools must be produced to achieve maximum efficiency, which may require large capital expenditures. If a single tool set is run through a die transfer system, it will have similar unit costs to a fixed die system.

Tool package design can be essentially the same for both types of systems, although specific details of the transport mechanism may necessitate the use of additional plates or mounting features. Design considerations for thermal transfer, component tolerances, and compression molding mechanics remain unchanged.

2.4 Tools for Low Temperature Applications

The engineering principles discussed above apply equally to chalcogenide materials as for oxide glasses. When molding chalcogenide glasses where temperature extremes are not as severe, other tooling materials may be considered. Characteristics such as thermal properties, mechanical strength, and the ability to create complex surfaces are primary considerations for these tools.

The use of composite materials in tool design is often employed to take advantage of specific requirements within the tool package. For example, a material may be easily diamond turned to create complex surfaces, but not possess the strength needed for a particular application. Designing non-homogeneous tool structures can often produce the desired result, and at much lower cost than for carbide tooling.

Transformation temperatures of chalcogenide materials are usually between 150 and 400 °C, which places them well below most oxide glasses. The relatively weak atomic bonds that define these glass types are responsible for their low Tg, and support consideration of tooling materials that may not be applicable for traditional oxide glass molding (Table 2.5).

Molding tools formed by these alternate materials may represent significant cost savings over tools made with carbide materials. Features that make these materials low cost are the relative ease of machining, low material cost, and resistance to thermal fatigue (prolonged thermal cycling).

A thorough comparison of tooling materials and overall project costs should be completed prior to the tool design phase. Several factors contribute to the overall tooling cost; basic mold tool costs, costs for resurfacing tools when worn or damaged, number of cycles between resurfacing, and number of allowable resurfacing operations before the tool needs to be replaced (Table 2.6).

As quantities increase, molding operations may take advantage of the ability for multi cavity molding. In this event, tooling costs increase, but overall lens costs decrease since cycle times remain relatively flat while output increases dramatically. As quantities increase, unit tooling costs reach an asymptotic low, and the overall lens cost approaches a minimum value (Fig. 2.17).

2.5 Temperature Sensing

Several options exist for temperature monitoring, and all come with their particular strengths and weaknesses. Thermocouples and optical pyrometers are perhaps the most common forms of thermal monitoring.

Thermocouples consist of a pair of metal wires made of dissimilar materials that are joined at one end (junction side) to read a voltage drop on the other end (measurement side). When the two ends are held at different temperature regimes, a voltage difference is created in the wires that can be measured and displayed. The amount of voltage difference is temperature dependent, so the system should be calibrated near the desired temperature to obtain accurate readings. Over time and with extended thermal cycling, the wires and casing can become brittle, causing failure of the thermocouple. Thermocouples are typically inexpensive and can be replaced at minimal cost.

Since the thermocouples do not actually touch the glass, it is imperative that they are located as near the glass as possible and that the end remain in contact with

Table 2.5 Common materials used in chalcogenide glass molding operations

	Density (ρ)	Cp (J/g K)	K (W/m K)	κ (m^2/s)	CTE (ppm/K)	Xm	ρ (Ω − m)
WC	15.40	0.20	82.0	2.66E−02	4.90E−06	1.00E−05	2.00E−08
Stavax	7.70	0.46	20.0	5.65E−03	1.11E−05	ferro	3.30E−06
CVD SiC	3.21	0.66	250.0	1.18E−01	4.50E−06	−1.28E−05	1.00E+00
a-Carbon	1.51	0.70	5.8	5.49E−03	2.10E−06	1.00E−04	3.50E−05
Ni	8.90	0.44	52.0	1.33E−02	1.30E−05	ferro	9.00E−08
Graphite	1.84	0.71	84.0	6.43E−02	6.00E−06	−6.00E−06	1.35E+01
Silicon (α)	2.33	0.70	13.8	8.46E−03	3.97E−06	−3.90E−06	3.50E+00
Invar	8.05	0.51	10.0	2.41E−03	7.80E−06	ferro	8.00E−07
Kovar	8.36	0.46	17.3	4.50E−03	5.06E−06	ferro	4.00E−07
Alum Alloy	2.95	0.85	115.0	4.59E−02	1.90E−05	1.65E−05	2.80E−08
Inconel	8.19	0.44	11.4	3.20E−03	1.30E−05	ferro	1.30E−06
Hastelloy	8.22	0.52	14.1	3.33E−03	1.42E−05	ferro	1.30E−06
Germanium	5.32	0.31	0.59	3.58E−04	6.30E−06	−7.68E−05	4.60E−01

Table 2.6 Exemplar tooling cost for various tooling materials and order quantities

Tooling cost per cavity. Assumes 1-2 cavity mold.					Overall Tool Costs for xx Units				
Material	Mold Tool Cost ($)	Refurb Cost ($)	Cycles Bet Refurb	Possible Refurb Cycles	10,000	25,000	50,000	75,000	100,000
WC or SiC	$7,000	$3,000	5000	5	$13,000	$22,000	$41,000	$60,000	$79,000
Composite	$4,000	$1,500	3000	3	$8,500	$21,000	$38,000	$56,500	$73,500
Low Cost	$2,500	$1,000	3000	5	$5,500	$12,000	$21,500	$33,500	$43,000
Other	$2,500	$1,000	3000	5	$5,500	$12,000	$21,500		

** Based on 1-2 cavity tooling, assumes identical support structures in place*

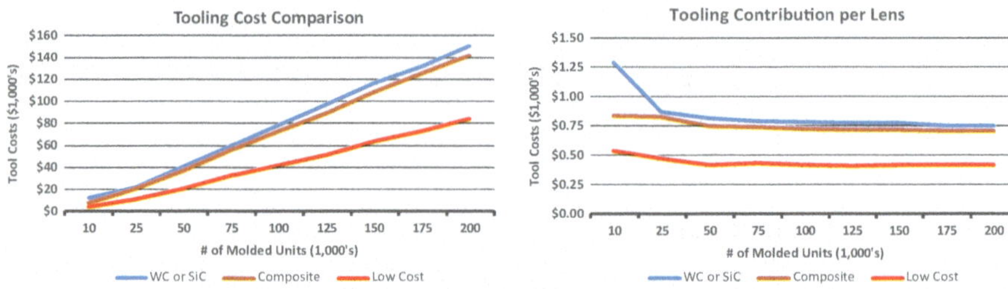

Fig. 2.17 Exemplar tooling cost and contribution per lens for various tooling materials

the desired material. A temperature difference will be present between the glass and thermocouple, and this delta must be considered when setting up the molding process.

Optical pyrometers operate by focusing thermal radiation from features in the molding system and onto a detector, which displays temperature readings based on the Stefan-Boltzmann Law. With these tools, the temperature of the glass preform surface may be read directly before pressing. These systems provide non-contact, accurate temperature readings but are highly dependent on the emissivity of the object. Emissivity is greatly affected by surface roughness, surface composition, and temperature of the surroundings, and can change based on the surface characteristics of the object under test and by gradual surface degradation.

For molding applications that operate in a closed chamber, the pyrometer must look through a glass enclosure to the object under test. With usage and continued thermal cycling, the clarity of the enclosure surface can be compromised, which will negatively affect the accuracy of the readings. These systems should be monitored and calibrated for accuracy on a regular basis.

2.6 Conclusion

The proper selection of tooling materials can have a profound effect on molding system performance and efficiency. Most systems will function quite well over some range of temperatures and molding conditions. However, when the molding environment requires large changes due to glass types or lens geometry, tooling designs should be reevaluated for best performance.

References

1. Fujilloy Co., Ltd., catalog 11–2007–TH, p 8–9.
2. SCHOTT TIE 32, Thermal loads on optical glass (SCHOTT AG, Aug 2004).

Chapter 3
Molding Surface Design and Useful Equations

A primary focus when designing the tool package is the management of thermal energy and thermal gradients that are a fundamental consequence of thermal transfer. Design choices will affect cycle time, optical accuracy, and geometric tolerances on the finished optic.

3.1 Thermal Compensation and the Molding Surface

In the previous chapter, we discussed the cause and effects of thermal transfer and thermal expansion in a mold tool system. We will now look more closely at the underlying reasons for these phenomena through the equations that govern their behavior.

The equations used to calculate changes in length of the tooling components can be applied to the molding surface to determine the correct profile needed to produce accurate optical surfaces. Just as the components expand and contract under the influence of temperature, the molding surface is subject to these same forces. Knowing when the glass will follow the tooling surface and when it will move on its own is the challenge of the design engineer.

This problem is of particular concern when working with mold tools that are made from anisotropic materials such as crystals. In this case, the surface expands and contracts at different rates based on the orientation of the material.

The mold dies, die holders, and certain other components are often manufactured of the same material. However, dissimilar materials may be used to improve heat transfer characteristics and to gain advantage for alignment between upper and lower molding surfaces. This is where differences in thermal expansion (CTE) parameters can work to the advantage of the designer. But first, it is useful to consider some basic principles of thermal energy transfer.

J. J. Nelson, *Precision Lens Molding of Glass: A Process Perspective*,
Progress in Optical Science and Photonics 8,
https://doi.org/10.1007/978-981-15-4238-1_3

3.2 Thermal Energy Transfer

Creation of thermal energy may be the result of one or more processes at work in the molding system. Infrared lamps are the most common type of heat source used in glass molding systems and rely on heat generation by radiation and absorption of electromagnetic energy. RF systems create an electrical current within the tool package and will therefore rely on excitation of electrons in the tool materials to generate heat. Resistive heating systems are similar to RF systems in that electrical current is used to initiate energy transfer, but the energy is generated outside of the tooling package. In each case, conduction is required to transfer thermal energy. Convection is also a component of the overall heat transfer in the system, but its main source of energy is thermal energy radiating from the tool package to the surrounding environment. This in turn serves to heat the outer surfaces of the preform and tool package.

3.2.1 Radiation and Absorption

Generation of thermal energy by infrared radiation involves absorption of incident radiation and subsequent conduction through the support tooling, molds, and glass preform. Radiation intensity follows the Inverse Square Law where intensity is proportional to $1/(\text{distance})^2$. The result of such a relation is that greater distances between heat source and tooling will result in longer cycle times. Best practice will always keep the IR source near to the substrate for maximum effectiveness; however, this would require different sets of IR lamps for each chamber size, which may be impractical and quite expensive. Some manufacturers of IR molding systems will supply different sized quartz tubes with their equipment to accommodate smaller diameter tool sets, but the energy fall off from the inverse square relation still holds.

Absorption of electromagnetic radiation occurs when energy from incident photons is transferred to atoms of the receiving material (substrate) and converted to thermal energy or another form of molecular energy. Increases in temperature reflect an increase in molecular motion of vibration in a material. As photons interact with atoms, their kinetic energy is transferred (absorbed), which causes an initial increase in vibrational amplitude. The quantum energy of infrared photons between 0.8 μm $\leq \lambda \geq 2.3\ \mu$m ($8.63 \times 10^{-20}$ J $\leq \Delta E \geq 2.48 \times 10^{-19}$ J) corresponds to the range of energies separating quantum states of molecular vibrations for many mold materials (Table 3.1).

Seeking a return to their lower energy state, the vibrating atoms release energy in the form of heat. This energy transfer is best understood through principles of quantum physics.

Table 3.1 Infrared energies for various spectral regions

Region	Wavelength (μm)	Energy (J)	Energy (eV)
Near IR	1–3	1.98E−19 to 6.62E−20	0.413 to 1.235
Mid IR	3–8	6.62E−20 to 2.48E−20	0.154 to 0.413
Far IR	8–50	2.48E−20 to 3.97E−21	0.024 to 0.154

Table 3.2 Lowest energy transition levels for various materials

	Germanium	Silicon	Tungsten Carbide
Energy bandgap at 300 K (eV)	0.66	1.12	0.47–0.87

Absorption occurs when the change in energy state of an atom relates to the energy of the photon according to (Table 3.2);

$$\Delta E = E_2 - E_1 = h\Delta v \tag{3.1}$$

h Planck's constant
ν Frequency of radiation

When no quantized energy levels in the substrate match the quantum energy of the incident radiation, the radiation will have little effect on the substrate. Absorption generally occurs only near the surface of the substrate since the intensity of the radiation decreases as discrete amounts of energy are absorbed by substrate atoms until an equivalent amount of energy is transferred between the photons and substrate. Higher energy photons have shorter wavelength and shallower absorption, while low energy photons possess longer wavelength and deeper absorption. Obviously, materials should be chosen that are efficient absorbers at the range of frequencies emitted by the source. The finite nature of this process is known as attenuation and establishes the effective depth of penetration for incident radiation as defined by the Beer-Lambert Law.

$$I(x) = I_0 e^{-\alpha \cdot x} \quad \alpha = \frac{1}{x} \ln \frac{I_0}{I} \tag{3.2}$$

I_0 Intensity of incoming light
x Distance to surface
α Absorption coefficient

The inverse of the absorption coefficient (1/α) equals the penetration depth of the incident radiation. As thermal energy is being created, some energy escapes the substrate to heat the surrounding atmosphere. This provides balance to the overall system and produces the energy needed for convection. The main source of energy for convection heating comes from heat being emitted from the support tooling and surrounding components.

3.2.2 Convection

In general, heat transfer between two objects of different temperatures follows Newton's Law, which states that the rate of temperature change in an object is proportional to the difference between the object temperature and the surrounding temperature. Stated another way, the temperature of an object at some future time will follow an exponential decay (or rise) and is dependent on the original conditions. Newton's Law is often stated as

$$\frac{d\Delta T}{dt} + \frac{1}{\tau}\Delta T = 0 \tag{3.3}$$

The solution to this first order differential equation can be written as

$$\Delta T(t) = \Delta T_o e^{-t/\tau} \tag{3.4}$$

The thermal time constant τ is a function of several material properties;

$$(\tau) = \rho C_p V / hAs \tag{3.5}$$

ρ Density (g/m^3)
h Heat Transfer Coefficient (W/m^2 K)
C_p Heat Capacity (J/g K)
A_S Surface Area (m^2)
V Volume of material (m^3)

This equation is most useful for analyzing the effects of convection heating and cooling when the molding process is performed in an inert, non-vacuum environment. In real life, convection heating and cooling of the molding system is much more complicated due to the input of fresh nitrogen and increased turbulence, but this equation will provide the engineer with a general idea of convection processes. More thorough analysis of cooling processes will likely require analytical software. However, while convection may be the primary process involved in cooling the tool package, it is just one of several at work in the heating cycle.

3.2.3 Conduction

Conduction processes are of great importance in the molding system. Not only is conduction the main source of temperature increase for the glass preform, but it also serves to manage thermal shock in the glass since energy transfer is limited by material properties in both glass and substrate. In cases where the design strength of the glass preform is less than the tensile stresses formed at the glass surface, steps must be taken to resolve this difference. Some possible solutions may be;

- Minimizing rapid temperature changes, especially during the cooling cycle
- Increasing preform strength through the polishing or removal of ground surfaces

Heat transfer equations based on Fourier's First Law relate specifically to conduction processes and describe the amount of heat conducted through a material per unit cross sectional area per unit time in the presence of a unit temperature gradient. Heat is defined as the energy transferred between a system and its surroundings due to a temperature difference between that system and some part of its surroundings. Therefore, the heat current is simply the flow of heat per unit time. The linear flow of heat in the x direction can be given by

$$H = -KA(dT/dx) \tag{3.6}$$

H Heat Current (J/s)
A Cross Sectional Area (m^2)
K Thermal Conductivity (W/m K)
T Temperature (K)

The negative sign indicates a positive flow of heat along the direction of decreasing temperatures. Considering Eqs. (3.4) and (3.6), it is obvious that when choosing tooling materials;

To increase heat flow and minimize process time, decrease;

- Density, Volume, Thickness in the direction of heat flow

To increase heat flow and minimize process time, increase;

- Heat Transfer Coefficient, Thermal Conductivity
- Ratio of Surface Area to Volume

The above discussions are helpful when examining one-dimensional heat transfer under steady state conditions. But once again actual experience is more complicated, and while the above statements are true and form the basis for thermal analysis, a more accurate solution will combine the many processes and consider their various interactions. In the PLM system, the rate of thermal energy input is often managed by a PID controller that adjusts energy input to avoid temperature overshoots and minimize cycle time, which results in a non-constant rate of energy input (Fig. 3.1).

The rate of heat conducted through a medium may vary with both position and time and has units of power per unit volume. The rate of heat conduction at position x_a minus the rate of conduction at position x_b equals the rate of change in energy within the substrate per unit time. It can be shown that Fourier's Law for transient heat conduction in one dimension is described by;

$$(\textit{Rectangular coordinates}) \quad \frac{\partial^2 T}{\partial x^2} = \frac{1}{\kappa}\left[\frac{\partial T}{\partial t}\right] \quad \text{Where } \kappa = \mathrm{K}/(\rho \cdot C_p) \tag{3.7}$$

$$(\textit{Cylindrical coordinates}) \quad \frac{1}{r}\frac{\partial}{\partial r}\left[r\frac{\partial T}{\partial r}\right] = \frac{1}{\kappa}\left[\frac{\partial T}{\partial t}\right] \tag{3.8}$$

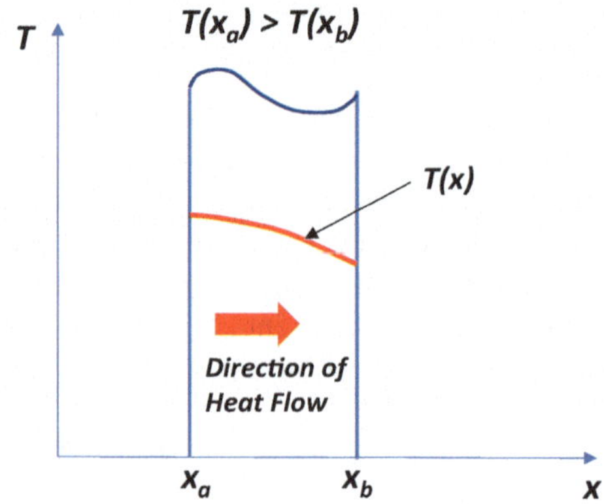

Fig. 3.1 Non-linear temperature distribution in a medium

and the symbol κ defines the thermal diffusivity of the material. The solution is found using separation of variables, with the thermal and spatial solutions written in Cartesian coordinates as;

$$T(t) = Ae^{-\kappa t/\lambda^2} \tag{3.9}$$

$$X(x) = B\cos\frac{x}{\lambda} + C\sin\frac{x}{\lambda} \tag{3.10}$$

and λ is a negative valued constant of separation (to ensure a finite solution) with linear units of measure. The general solution is then;

$$U(x,t) = T(t)X(x) \tag{3.11}$$

$$U(x,t) = Ae^{-\kappa t/\lambda^2}\left[B\cos\frac{x}{\lambda} + C\sin\frac{x}{\lambda}\right] \tag{3.12}$$

$$U(x,t) = e^{-\kappa t/\lambda^2}\left[D\cos\frac{x}{\lambda} + E\sin\frac{x}{\lambda}\right] \tag{3.13}$$

If we assume boundary conditions such that the initial temperature $T_{t=0} = 0$ at both ends of the conducting medium, then the general solution becomes;

$$U_n(x,t) = E_n e^{-\kappa(n\pi/L)^2 t}\left[\sin\frac{n\pi x}{L}\right] \tag{3.14}$$

When heat is generated within the system, as in the case of RF induction heaters, Fourier's Law for transient heat conduction in one dimension can be described by;

$$(Rectangular\ coordinates)\quad \frac{\partial^2 T}{\partial x^2} + E_{Gen} = \frac{1}{\kappa}\left[\frac{\partial T}{\partial t}\right] \tag{3.15}$$

$$(Cylindrical\ coordinates)\quad \frac{1}{r}\frac{\partial}{\partial r}\left[r\frac{\partial T}{\partial r}\right] + E_{Gen} = \frac{1}{\kappa}\left[\frac{\partial T}{\partial t}\right] \tag{3.16}$$

A similar solution can be defined for these equations using the same methods of separation of variables.

Without exception, the transfer of thermal energy and management of thermal gradients must be a primary focus for producing accurate surfaces when using PLM processes. In most cases, we are concerned with the initial and final thermal states of the tool package and assume no problems in between. This philosophy holds minimal risks and saves analysis time; however, it is always a good idea to perform a thorough investigation at least once in order to develop a better understanding of thermal processes.

Thermal gradients occur in the direction of heat flow, and the goal of every designer is the management (usually the reduction) of thermal gradients. To minimize cycle times and reduce process costs, the focus is set on conducting thermal energy both into and out of the tool package. Material choices and component geometry become the most common devices used by the designer to achieve their goal.

The presence of thermal gradients are particularly troublesome when working with "short glass" types. We define "short glasses" as those which demonstrate a rapid change in viscosity with unit change in temperature (large $\Delta\eta/\Delta T$), which results in a very narrow working range. On the contrary, "long glasses" demonstrate a slow or gradual change in viscosity with respect to temperature (small $\Delta\eta/\Delta T$) and therefore have a much larger process window (Fig. 3.2).

Looking back on Eqs. (3.4) and (3.6), it is easily seen that heat flow is a function of the temperature difference between two bodies, and the greater flow occurs when the temperature difference is greatest. This is important to remember when working with short glasses. Since viscosity of these glasses is greatly affected by temperature, care must be taken to ensure that a proper soak time is allowed for even temperature distribution across the preform before pressing. Temperature rise may be relatively slow towards the end of the cycle, but adequate time must be allowed for complete transfer to occur.

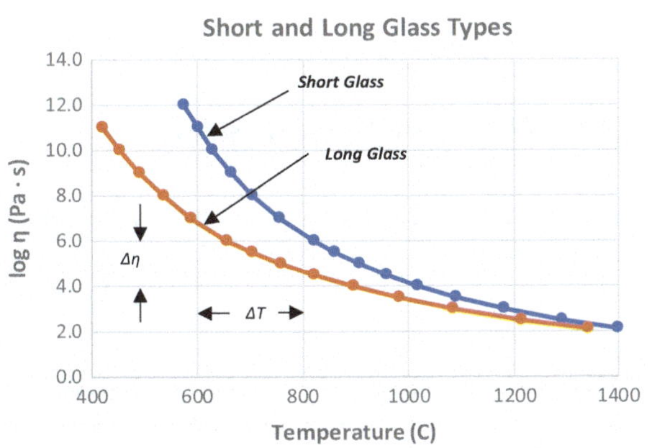

Fig. 3.2 Sample viscosity curves for long and short glasses

3.3 Thermal Expansion of Tooling

The basic equation that governs linear changes in a material due to varying thermal conditions is

$$\mathrm{L_f} = \mathrm{L_o}(1 + \alpha\,\Delta\mathrm{T}) = \mathrm{L_o} + \Delta\mathrm{L} \tag{3.17}$$

$\mathrm{L_f}$ = Final dimension of component
$\mathrm{L_o}$ = Original dimension of component
$\Delta\mathrm{T} = (\mathrm{T_f} - \mathrm{T_o})$ in units K or °C
$\Delta\mathrm{L}$ = Linear strain = $(\mathrm{L_o}\,\alpha\,\Delta\mathrm{T})$ in linear units
α = Linear coefficient of thermal expansion (usually as ppm/K or ppm/°C).

This equation is commonly used to calculate the tool surface figure for molding applications. Since the tool material and glass expand and contract at different rates, the surface that is cut into the tool is generally not the same as the lens prescription. It can be shown that for first order approximation, the surface of the tool must be modified in the following manner to produce the correct lens surface;

$$\mathrm{L_{RT(T)}} = \mathrm{L_{RT(L)}}\left(1 - \alpha_\mathrm{T}(\mathrm{T_P} - \mathrm{RT}) + \alpha_\mathrm{L}(\mathrm{T_P} - \mathrm{RT}) - \alpha_\mathrm{L}\alpha_\mathrm{T}(\mathrm{T_P} - \mathrm{RT})^2\right) \tag{3.18}$$

For applications where low processing temperatures exist, this equation may be simplified to

$$\mathrm{L_{RT(T)}} = \mathrm{L_{RT(L)}}(1 + (\alpha_\mathrm{L} - \alpha_\mathrm{T})(\mathrm{T_P} - \mathrm{RT})) \tag{3.19}$$

$\mathrm{L_{RT(T)}}$ Dimension of the tool at room temperature
$\mathrm{L_{RT(L)}}$ Dimension of the lens at room temperature
α_L Linear coefficient of thermal expansion of the lens (ppm/K or ppm/°C)
α_T Linear coefficient of thermal expansion of the tool (ppm/K or ppm/°C)
$\mathrm{T_P}$ Process temperature, (K or °C)
RT Room temperature, (K or °C)

3.3.1 Thermally Induced Stress

From Eq. (2.1), thermally induced stress in the glass can be written as;

$$\sigma_{th} = f \cdot \frac{(\alpha \cdot E)}{(1 - \mu)} \cdot \Delta T \tag{3.20}$$

The glass specific factor "*f*" should now be considered for this equation. The worst case scenario is that in which the glass experiences an extremely rapid temperature change and thermal stress is at its maximum. In this case, temperature change is faster than conduction will allow, and both ΔT and ΔL between the outer skin of the glass and regions beneath the exterior are at their greatest. Use a value of $f = 1$.

When temperature change is moderate and thermal conduction is able to mitigate some temperature difference in the glass, use a value of $0.5 \le f \ge 0.7$. The lowest stress condition is when temperature change occurs slowly and conduction processes are allowed to fully moderate potential temperature differences. In this case, use a value of $f \le 0.5$.

The value calculated for thermal stress should always be compared against the allowable stress of the material and a suitable safety factor applied to determine whether the thermal impact will be greater than the material can withstand. The difficulty with this approach is that accurate information for many material properties is not readily available for elevated temperatures. Therefore, as much information as possible should be obtained in the temperature regime of interest before a reasonable estimate can be made.

3.3.2 *Stress Induced Wavefront Distortion*

Internal stress in glass can cause birefringence, and as a result, an unacceptable level of wavefront distortion may be evidenced in the molded article. The following equation may be used to estimate wavefront distortion due to internal stress in the glass;

$$\Delta W_p = K\sigma(t/\lambda) \tag{3.21}$$

ΔW_p Wavefront retardance between principal polarizations (waves)
K Stress optic coefficient (mm^2/N)
σ Difference between minimum and maximum principal stress in the glass
t Thickness of the glass region examined
λ Wavelength of light

The stress optic coefficient for some common moldable glass types is shown in Table 3.3;

Table 3.3 Stress optic coefficient of some moldable glasses

Glass manufacturer	Glass type	Stress optic coefficient
SCHOTT	P-BK7	2.77
SCHOTT	P-LAF37	2.26
SCHOTT	P-LAK35	1.76
SCHOTT	P-PK53	2.06
SCHOTT	P-SF8	2.73
SCHOTT	P-SF68	1.61

3.4 Tool Design for Isotropic Materials

For all applications of molding technology, thermal expansion of components plays a key role in the design of the tooling package. While some molding houses may choose to make all components of the same material to avoid issues with clearance and fits while pressing, they miss out on the opportunity to improve the accuracy of the finished molded lens. The best designs will consider the physical and thermal properties of all materials when forming their solution.

When designing optics that are manufactured by forming glass at elevated temperatures, it is necessary to accommodate for thermal expansion differences between the optical materials and mold materials. Typically, this is accomplished by adjusting the constants from the asphere equation of the design surface to compensate for thermal expansion of the optical material as it is heated from its design temperature to the processing temperature. These coefficients are then adjusted for the thermal contraction of the mold material as it is cooled to standard temperature, considering also the temperature range wherein the glass remains formable.

When working with oxide glasses and isotropic materials such as carbides, the solution is rather straightforward with application of the above noted equations. However, when working with chalcogenide materials and some low Tg oxide glasses, the opportunity exists for using lower cost tooling materials that provide machining options and the potential of molding unique shapes and features. Some of these materials have a crystalline structure or have been formed by methods that cause them to behave with anisotropic properties. The application of these equations and the solution for anisotropic conditions should apply equally to that of isotropic materials and is the proof that the solutions are valid.

3.5 Tool Design for Non-isotropic Materials [1]

Glass molding technology has historically known the benefits of working with materials that possess isotropic thermal expansion. Certainly, this is the case for glasses, but it may not be the case for all mold materials. The purpose of this section is to explore how adjustment of the constants of the asphere equation are altered by anisotropic thermal expansion of the mold material. Isotropy of the lens material is still assumed.

Anisotropy of material properties can arise from the fundamental crystallographic characteristic of the material or from the means of production. Typically, the thermal expansion of bulk materials can be characterized by measuring the expansion along two orthogonal directions. Thus, it is necessary to ensure that one of these directions is parallel to the optical axis of the mold that is being constructed. Deviation from this case will result in astigmatism of the optical surface as it is heated or cooled from its fabrication temperature.

A simple thought experiment reveals one of the differences from the isotropic mediums that must be dealt with. Consider a sphere that is made from a material with different thermal expansion along orthogonal directions. If the sphere is heated, it will elongate more along one axis than the other. That is, the sphere has now become an ellipse.

This indicates that unlike isotropic expansion, we must anticipate a change of vertex radius as well as a change in eccentricity of the conic term ($k = -\text{eccentricity}^2$). Since spheres are just special cases of the more generalized asphere equation, we will use the asphere equation for analysis and then examine the special case of spheres. The impact of the conic term will be examined first, followed by the higher order terms.

3.6 Design of the Molding Surface

3.6.1 Spherical and Aspheric Surfaces

Conic Term:

The following terms will be used for the analysis:

L Length
L* Length at elevated temperature
x_{exp} Expansion in the x direction
y_{exp} Expansion in the y direction

Thus,

$$L_x^* = L(1 + \alpha_x \Delta T) = L(x_{exp}),\ \text{and}$$
$$L_y^* = L(1 + \alpha_y \Delta T) = L(y_{exp}) \tag{3.22}$$

The conic portion of the standard form of the general aspheric equation can be written as

$$X = SAG = \frac{Y^2/R}{1 + \surd(1 - (k+1)Y^2/R^2)} \tag{3.23}$$

After rearranging terms, the implicit expression of a conic with its vertex at the origin and symmetry along the x axis can be given as:

$$Y^2 - 2RX + (k+1)X^2 = 0 \tag{3.24}$$

Applying thermal expansion to the X and Y dimensions and anticipating a change of R and k, we can write:

$$Y^2(y_{exp})^2 - 2R^*X(x_{exp}) + (k^*+1)X^2(x_{exp})^2 = 0 \tag{3.25}$$

From this expression we can write the solution:

$$R^* = \frac{R(yexp)^2}{(xexp)} \quad \text{and} \quad k^* = \frac{(k+1)(yexp)^2}{(xexp)^2} - 1$$
$$C^* = 1/R^* = \frac{(xexp)}{R\ (yexp)^2} = \frac{C(xexp)}{(yexp)^2} \tag{3.26}$$

For the case of thermal contraction, it is obvious that

$$R = \frac{R*(xexp)}{(yexp)^2} \quad \text{and} \quad k = \frac{(k*+1)(xexp)^2}{(yexp)^2} - 1$$
$$C = \frac{C*(yexp)^2}{(xexp)} \tag{3.27}$$

As we would hope for the isotropic case ($x_{exp} = y_{exp}$), the above reduces to the standard forms

$$R^* = R(exp) \quad C^* = C(exp) \quad k^* = k \tag{3.28}$$

Expansion Terms:

The higher order expansion terms of the aspheric equation may be generalized to the form

$$X_n = P_n(Y^n) \tag{3.29}$$

Allowing for thermal expansion and anticipating a change in P we can write

$$X_n(x_{exp}) = P_n(Y^n)(y_{exp})^n \tag{3.30}$$

$$P_n(Y^n)(x_{exp}) = P_n(Y^n)(y_{exp})^n \tag{3.31}$$

$$P_n^* = P_n(x_{exp})/(y_{exp})^n \tag{3.32}$$

And in the case of thermal contraction, it is again obvious that

$$P_n = P_n^*(y_{exp})^n/(x_{exp}) \tag{3.33}$$

Again, for the case of isotropic expansion, these reduce to the usual equations

$$P_n^* = P_n/(exp)^{(n-1)} \quad \text{and} \quad P_n = P_n^*(exp)^{(n-1)} \tag{3.34}$$

Table 3.4 Sample calculations of aspheric molding surfaces

SURFACE 2	Lens Surface 2	Tool Surface 2
Enter Clear Aperture	23.800	23.858039
Enter "C" Coefficient (1/Radius)	0.011039	0.011012
Enter "k" Value	0.000000	0.000000
Enter "D" Coefficient (4th Power)	-2.48246E-06	-2.46439E-06
Enter "E" Coefficient (6th Power)	-1.76972E-09	-1.74830E-09
Enter "F" Coefficient (8th Power)	-1.75209E-12	-1.72247E-12
Enter "G" Coefficient (10th Power)	0.00000E+00	0.00000E+00
Enter "H" Coefficient (12th Power)	0.00000E+00	0.00000E+00
Enter "J" Coefficient (14th Power)	0.00000E+00	0.00000E+00
Enter "K" Coefficient (16th Power)	0.00000E+00	0.00000E+00
Enter "L" Coefficient (18th Power)	0.00000E+00	0.00000E+00
Enter "M" Coefficient (20th Power)	0.00000E+00	0.00000E+00
Enter Preform Radius	6.037	
Lens Base Radius (Calculated)	90.5900	
Tool Base Radius (Calculated)		90.81091

Diffractive Surfaces	Lens Surface 2	Tool Surface 2
Enter design wavelength (mm)	1.00000E-02	
Enter index @ λ_o (mm)	2.76608E+00	
Enter 1st phase coefficient (ρ^2)	-3.15636E-04	-3.14868E-04
Enter 2nd phase coefficient (ρ^4)		0.00000E+00
Enter 3rd phase coefficient (ρ^6)		0.00000E+00
Enter 4th phase coefficient (ρ^8)		0.00000E+00
Enter 5th phase coefficient (ρ^{10})		0.00000E+00
Calculate Step Height (mm)	5.66226E-03	5.67607E-03
Diffractive Zones		
Enter zone radius for ring 1	5.629	5.6427
Enter zone radius for ring 2	7.960	7.9794
Enter zone radius for ring 3	9.749	9.7728
Enter zone radius for ring 4	11.257	11.2845
Enter zone radius for ring 5		0.0000

3.6.2 Diffractive Terms

The accurate replication of diffractive surfaces requires similar consideration of thermal expansion for both the tool surface and glass article. Diffractive surfaces are normally designated with values for zone height and radial distance. Each of these features follows the same rules shown above for expansion and contraction. Application of the rules shown above will accurately define the room temperature tool surface that is needed to create the finished glass article.

3.6.3 Putting It All Together

Table 3.4 shows an example of sample calculations for a system using polycrystalline germanium molding tools and IRG-26 infrared glass. The final surface of the molded lens is known and entered into the chart; the dimensions of the tool surface at room temperature are calculated values.

In this case, thermal expansion of the glass is greater than that of the tool material. Therefore, dimensions of the tool surface at room temperature will always be greater than the final lens surface dimensions since the glass changes shape faster than the tool surface after pressing.

3.7 Conclusion

The proper selection of tooling materials and coatings can have a profound effect on molding system performance and efficiency. Most systems will function quite well over some range of temperatures and molding conditions. However, when the molding environment requires large changes due to glass types or lens geometry, tooling designs should be reevaluated for best performance.

Reference

1. "Tool Design for Non-Isotropic Materials" based on discussions with Dr. John Pulver, Eastman Kodak Company

Chapter 4
Tool Coatings

Tool coatings are perhaps the most important and most underappreciated component of the tool package—they are the strong protector of the precision molding surface, with the endurance of a marathon runner, yet are almost unnoticeable for size.

4.1 Chemical Interactions

The manufacture of thin film coatings is not at all analogous to the application of amorphous materials on a surface where asperities are filled and irregularities are smoothed to form a perfect surface, free from defects and irregularities. Rather, thin film coatings are created by the building up of atom upon atom, layer upon layer, so that the substrate surface is replicated as near perfect as possible. Therefore, substrates should possess surface figure at least as good as the desired final shape, and materials must be chosen to ensure good adhesion and longevity.

A basic understanding of chemistry is vital to the design of thin film coatings. Coatings used in the glass molding process can be categorized into two main types; protective coatings for the molding tools, and release coatings for the freshly molded glass. Since most glasses that have favorable properties in the visible range of the electromagnetic spectrum contain oxygen, and oxides are commonly formed on the surface of many materials, let us consider an oxygen atom and the mechanisms involved for combining with other elements.

Diatomic oxygen (O_2) forms a double covalent bond with itself, meaning that two pairs of electrons are shared by the two oxygen atoms. The bonds formed by these two pairs are quite strong. This sharing occurs at the 2p shell, which leaves openings for additional electrons to fill out the shell. It is in these vacancies that other materials seek to combine with the O_2. As thermal energy is added to the system (as done in glass molding), atoms gain mobility and additional degrees of freedom, which increases the probability of grouping with other elements.

J. J. Nelson, *Precision Lens Molding of Glass: A Process Perspective*,
Progress in Optical Science and Photonics 8,
https://doi.org/10.1007/978-981-15-4238-1_4

A single oxygen atom has 8 electrons; 2 electrons reside in the 1st level s subshell, 2 in the 2nd level s subshell, and 4 in the 2nd level p subshell. Since the 2nd level p subshell can accommodate a total of 6 electrons, this leaves 2 openings for electronic bonding with other materials. When combining with other elements, electrons generally fill into open energy levels (shells numbered 1, 2, 3, …), subshells (labelled s, p, d, etc.), and orbitals to produce the lowest possible energy arrangement according to the Aufbau Principle (Fig. 4.1).

So, what are the electrons of these oxides looking for? Very simply, they are looking to combine with other materials and settle into a lower energy state since a thermodynamic force is always present to minimize surface and interfacial energies. Surface atoms have lower coordination numbers than their counterparts in the substrate, and therefore possess greater energy levels (as much as 20–30% greater). Oxygen has unbound electrons in the 2p shell, so the nearest energy move is in combination with the 3d shell of a neighboring element. To resist this interaction, materials that possess a filled 3d shell find good application as release coatings for glass molding (Fig. 4.2).

Certainly, there are more elements to consider than simply oxygen, but we use oxygen as an example of how elements combine to settle on a stable energy configuration. Physical interactions also play a role in the adhesion of thin film coatings since rougher surfaces provide greater surface area and the ability for mechanical "grab" to resist delamination, but it is assumed that smooth surfaces are encountered where surface roughness is minimized.

As thermal energy is added to the system, vibrational amplitude of the elements is increased as well as their surface mobility and ability to actively join with other materials. This increase of available energy helps to produce a strong bond between the substrate and materials commonly applied by thin film coating processes. However, as materials return to room temperature, surface stress is increased due to thermal expansion mismatch between materials.

In general, noble materials are sought after as both protective and release coatings for glass molding operations. In a strict definition of noble materials according to physics, noble metals are those that have a filled electron d-band (copper, silver, and gold). We place a second condition on our group of materials to include those

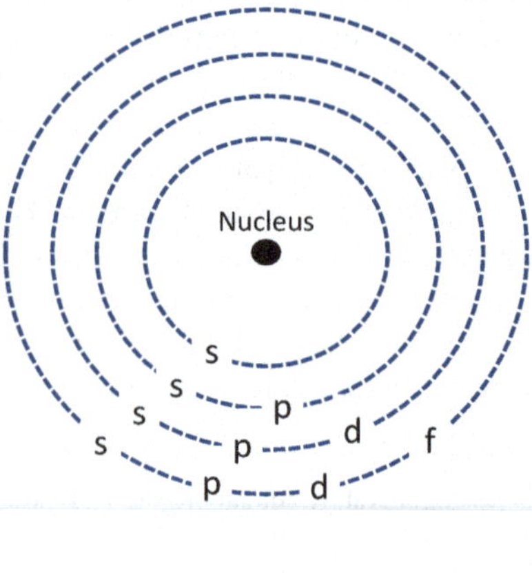

Fig. 4.1 Exemplar image of shells and subshells

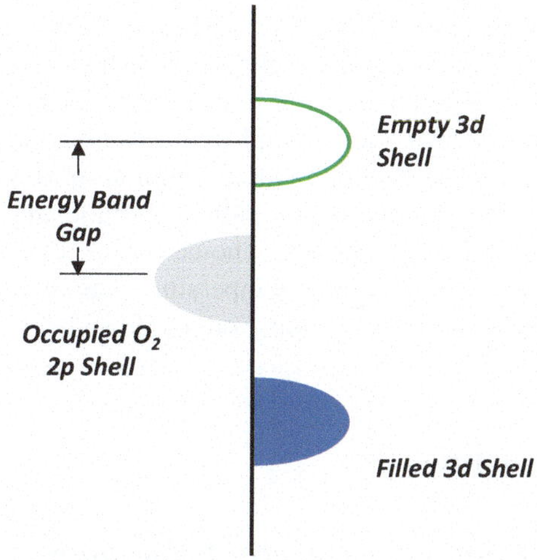

Fig. 4.2 Illustration of shells and energy levels

that resist oxidation and corrosion, even when heated. Copper for instance, has the property of a filled d-band but does not resist oxidation. For our purposes, we consider materials such as silver, gold, platinum, rhodium, palladium, rhenium, osmium, and iridium. These additional materials do not possess completely filled d bands, but are typically very stable and non-reactive over a wide temperature range. The opposite of a noble metal is a base metal.

The term "noble metal" may also be used to describe its chemical and galvanic activity, and materials may be ranked according to whether they are more noble or more active. However, in this context, graphite (a form of carbon) is more noble than many noble metals (see Fig. 4.3), and as such, is often used as both a protective and release coating in glass molding operations.

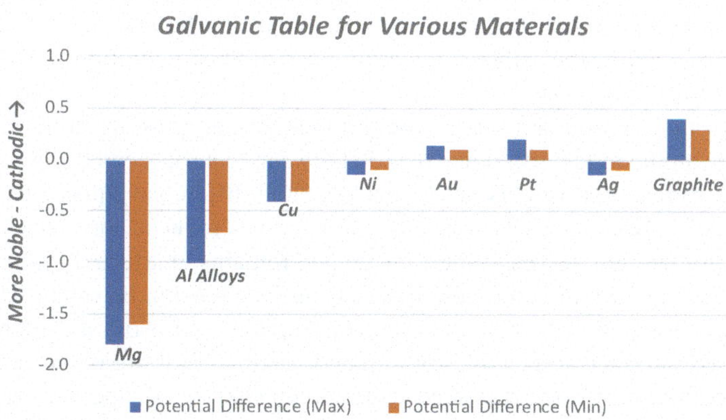

Fig. 4.3 Electric potentials of various materials

While the coating needs to act as a release surface to the heated glass, it must also possess properties that allow it to adhere well to the substrate and protect it from damage. As noted in Chap. 2, different materials may be used for substrates that optimize performance based on the heat source employed, machinability, durability, or the ability to withstand thermal degradation. Some common substrates for oxide glasses are members of the carbide family such as silicon carbide and binderless tungsten carbide. High phosphor nickel is a popular choice for use with chalcogenide materials since they possess a lower molding temperature than their oxide counterparts. Vitreous carbon has found application with both oxide and chalcogenide glasses.

4.2 Protective Coatings

4.2.1 Protective Coatings—General Information

A partial list of necessary attributes for protective coatings includes:

- Good adhesion to the substrate
- The ability to resist corrosion between coating and substrate
- Low wear from sliding forces
- Sufficient hardness to resist scratches
- Good resistance to both compressive and shear forces
- Good resistance to degradation due to thermal cycling.

Material compatibility between substrate and coating elements is often overlooked regarding the potential for degradation due to intrinsic electric currents, or galvanic corrosion. Galvanic corrosion occurs when dissimilar metals contact each other and cause oxidation or corrosion. For this reaction to take place, the materials must have an electrochemical difference between them and an electrically conductive path for current to flow. This difference in potential provides energy for the flow of electric current. A conductive path for the metal ions must also be present to enable flow from the anodic and to the cathodic material. Each of these conditions must exist for galvanic corrosion to occur.

For controlled environments like those common to many glass molding operations, a maximum anodic index difference of 0.50 between materials can be allowed without significant reaction. In general, higher potential difference equates to a greater probability of galvanic activity as the anode becomes reduced, or corroded, while the nobler of materials serves as the cathode and remains unaffected (Table 4.1).

Surface energy of a material, and likewise, surface tension between two dissimilar materials, may contribute to coating delamination, and is often evidenced by a compromised interface at the edges of tool and coating. Tool life is shortened when protective coatings fail since damaged coatings can also lead to openings such as pin holes and pull outs that expose the substrate to the softened glass, which is then able

Table 4.1 Anodic Index of various materials [5]

Material category	Anodic index
Gold and Au–Pt alloys	0.00
Rhodium plating	0.05
Silver, high Ni–Cu alloys	0.15
Nickel, Titanium	0.30
Copper	0.35
Brass, Bronze	0.40
Aluminum	0.75

to combine with the elements of the tool and cause significant damage. Maintaining consistent surface quality across all coated surfaces is fundamental to minimize differences in surface tension.

Thermal expansion differences between substrate and coating should be considered since this is a major factor in determining coating stress, and high coating stress increases the probability of coating delamination and failure. Differences in thermal expansion do not need to exactly match, but they should be minimized. Assuming a linear change in CTE mismatch over the temperature range and isotropic behavior from the substrate, the thermal stress generated at the coating-substrate interface can be described by Eq. 4.1,

$$\sigma(T) = \int_{To}^{Tf} \Delta\alpha G dT = \frac{E}{(1-\nu)} \Delta\bar{\alpha}\Delta T \tag{4.1}$$

$\Delta\alpha$ difference in CTE values
G shear modulus factor
E Young's modulus
$\Delta\bar{\alpha}$ average difference in CTE values
ΔT temperature range
ν Poisson's ratio.

This is not to be confused with the expansion mismatch discussed in the previous chapter (Chap. 3, Sects. 3.3–3.6) that explains dimensional changes due to changes in temperature, or stress formed in the glass due to cooling. Thermal expansion coefficients for some materials are shown in Table 4.2 for a temperature range of 20 °C $\leq$ T $\geq$ 300 °C.

4.2.2 Protective Coatings for Oxide Glass Applications

Fortunately, a few materials have demonstrated superior performance and satisfy many of these requirements, thereby reducing the need for coating designs that are specific for every glass type or substrate. The most common protective coating applied

Table 4.2 Thermal expansion coefficients of various materials

	CTE (ppm/K)
Substrate materials	
WC	4.90E−06
CVD SiC	4.50E−06
α-Carbon	2.10E−06
Ni	1.30E−05
Silicon (α)	3.97E−06
Graphite	6.00E−06
RSA Alum	1.90E−05
Germanium	5.70E−06
Coating materials	
Pt	9.00E−06
Pt−Ir	8.70E−06
DLC	1.00E−06
α-Carbon	2.10E−06
Graphite	6.00E−06
Gold	1.42E−05
TiN	9.35E−06
SiN	2.80E−06
Oxide glasses	
P-BK7	7.30E−06
P-SF8	1.11E−05
P-LAK35	9.70E−06
P-SF68	9.70E−06
L-BAL35	8.10E−06
L-LAH53	7.20E−06
L-PHL1	1.40E−05
L-TIM28	1.30E−05
Chalcogenide glasses	
$Ge_{22}As_{12}Se_{55}$	1.21E−05
$Ge_{30}As_{13}Se_{32}Te$	1.34E−05
$Ge_{10}As_{40}Se_{50}$	2.04E−05
$Ge_{28}Sb_{12}Se_{60}$	1.40E−05
$As_{40}Se_{60}$	2.08E−05

to tools for oxide glass molding is a blend of platinum and iridium, typically consisting of around 80% platinum. These coatings not only demonstrate good durability when exposed to a wide range of oxide glasses, but also provide good release from the glass upon cooling.

The use of platinum based materials in the glass making industry may have begun as early as 1829 when Michael Faraday conducted experiments melting glass in containers of pure platinum [1]. Faraday demonstrated one of the main benefits from using platinum to process glass is the exceptionally high glass purity that results from melting since platinum retains its noble properties even at highly elevated temperatures. Since that time, platinum (Pt) and its alloys have been used extensively in the melting of oxide glasses. Their high melting point, superior resistance to oxidation at glass processing temperatures, and virtual insolubility with nearly all oxide glasses makes them a solid choice for a variety of applications. These properties, when coupled with the high quality of glass produced, certainly offset platinum's high initial cost when high volumes of high quality glass articles are required.

Platinum may be combined with rhodium (Rh) to form a material that is resistant to even higher temperatures than pure platinum. Platinum and its alloys do not form oxides and therefore, protect the tool surface from the very active oxide compounds found in glass, especially at highly elevated temperatures. Another benefit is that the coefficient of thermal expansion (CTE) of platinum is very near that of many oxide glasses (9.0 ppm/°C for $20\ ^\circ\text{C} \leq T \geq 1000\ ^\circ\text{C}$).

Platinum may also be combined with iridium (Ir), which is a very hard and dense material that provides improved hardness and scratch resistance. Mixtures of platinum and iridium are often found as tool coatings since they excel as both a protective coating and a release coating. Longevity of Pt–Ir coatings can exceed 3000 molding cycles, but the material is very expensive and significantly increases tooling costs. For this reason, Pt based coatings are usually consigned to high volume, high precision applications.

Obviously, carbon based coatings such as diamond like carbon (DLC) and carbon graphite have a natural affinity with carbide based substrates. These materials will provide good adhesion and may offer good protection, depending on the application.

Graphite has proven to be a useful tool coating for low volume molding applications of oxide glasses when applied with physical vapor deposition (PVD) or chemical vapor deposition (CVD) systems. Application by these methods will provide the strongest possible bond to the substrate while creating a smooth coated surface with nearly perfect replication of the substrate.

Carbon has an electronic arrangement of $1s^2 2s^2 2p^2$ with four valence electrons in the 2s and 2p orbitals. The most stable bonding configuration of carbon under normal environmental conditions is graphite, where the s orbitals mix with only one of the two p orbitals, forming three sp^2 hybrids. Graphite is cathodic (see Fig. 4.3) and naturally bonds with carbide based materials, but layers of the graphite crystal are held together by weak Van der Waals forces that create a "softness" to the structure. So even though graphite has a natural adhesion to carbide substrates and inhibits adhesion with many oxide glasses, its structure is weak and may be quickly removed by sliding forces encountered when the heated glass moves across the tool surface.

Table 4.3 Electronic configuration of various noble materials

Element	Symbol	e-configuration
Ruthenium	Ru	$[Kr]4d^7 5s^1$
Rhodium	Rh	$[Kr]4d^8 5s1$
Palladium	Pd	$[Kr]4d^{10}$
Silver	Ag	$[Kr]4d^{10} 5s^1$
Rhenium	Re	$[Xe]4f^{14} 5d^5 6s^2$
Osmium	Os	$[Xe]4f^{14} 5d^6 6s^2$
Iridium	Ir	$[Xe]4f^{14} 5d^7 6s^2$
Platinum	Pt	$[Xe]4f^{14} 5d^9 6s^1$
Gold	Au	$[Xe]4f^{14} 5d^{10} 6s^1$
Graphite	C	$1s^2 2s^2 2p^2$

Therefore, graphite is most often used as a tool coating for low volume applications (Table 4.3).

Amorphous hydrogenous carbon, commonly known as diamond like carbon (DLC), has gained popularity in recent years as a tool coating for a variety of glass molding applications. DLC coatings contain the same building blocks as graphite coatings, but their solid structure results in a coating with very high hardness. In order to form the stronger covalent bonds, the s orbitals mix with three p orbitals to form a very strong sp^3 hybrid tetrahedral structure. Diamond is about 60% denser than graphite, and with all of the electronic bonds having the same length and angle, the resultant structure is highly uniform and of consistent strength in all directions.

Diamond like carbon coatings can be applied with either PVD or CVD systems, each having certain benefits. In PVD processes, atoms from the coating material are deposited by thermal or mechanical energetic impact to the surface of the substrate while under a controlled vacuum environment. This produces coatings of high purity, mobility, and energy, with the resultant coating exhibiting high density and uniformity. CVD processes occur by depositing elements of chemically reactive vapors onto the substrate. In CVD processes, coating is applied wherever the vapors come in contact with the substrate and may be performed at atmospheric pressure. CVD coatings are often considered when uniform layers are needed across complex surface geometries.

So how can diamond and graphite be so different when they consist of the same building blocks of carbon? It is because of the manner in which the carbon atoms are arranged in each material. Both materials are considered crystalline, but diamond has a three dimensional tetrahedral structure where each carbon atom is joined to four other carbon atoms with a high degree of symmetry. Each bond length is identical (1.545 Å) and at an angle of 109.5°. This gives the material very high resistance to compressive forces and extremely high hardness (measured as 10 on the Mohs scale, making diamond the hardest material on earth). This consistently regular structure also helps make diamond the best conductor of thermal energy while being one of the best electrical insulators (Table 4.4).

Table 4.4 Select properties of graphite and diamond

	Graphite	Diamond
2d bond length (Å)	1.421	1.545
3d bond length (Å)	3.347	1.545
Bond angle (°)	120	109.5
Density (g/cm^3)	2.26	3.51
Coordination number	3	4
Hardness (Mohs Scale)	1	10
Thermal conductivity (W/m K)	102	2200
Electrical resistivity (μ cm)	1270	$\approx 10^{20}$

The carbon atoms in graphite are also arranged as an infinite array, but form a layered structure where each carbon atom is joined to three other carbon atoms in a hexagonal structure at an angle of 120°. Spacing between layers is greater than the spacing between atoms within the hexagonal (3.347 Å vs. 1.421 Å), and therefore lacks the symmetry found in diamond. These extended bonds between layers produce a lower density for graphite than diamond (2.26 g/cm^3 vs. 3.51 g/cm^3), while the longer spacing between layers also makes graphite prone to shearing forces.

Amorphous carbon, vitreous carbon (VC), and glassy carbon are all names for the same basic material that is yet another variation on carbon based tool surfacing. This material combines both glassy (amorphous) and crystalline properties that results in a fullerene related microstructure that does not suffer from concerns over grain size as do many materials used for oxide molding (ref Sect. 2.1.1). It is a low density material with good thermal conductivity and fair resistance to mechanical loading but may fracture under high loads due to its glassy structure. When used as a coating material, it will replicate substrate features with high fidelity. While this material can be polished in a manner similar to oxide glass, it is difficult to machine by diamond turning methods, which may limit its utility for some molding applications.

Titanium nitride (TiN), titanium aluminum nitride (TiAlN), and boron nitride (BN) are also used as low cost coating materials more commonly found in machine tools. However, they have shown some value for very specific molding applications, usually when low temperature, low pressing force, and low quality surface is required. Coatings made with these materials do not generally produce high quality optical surfaces due to their chemical basis and structure and may not yield acceptable results for precision molded optics with application in the visible portion of the spectrum.

4.2.3 *Protective Coatings for Chalcogenide Glass Applications*

Chalcogenide materials differ from oxides in their basic structure and chemistry, and many of the above mentioned principles no longer apply. While oxygen is an integral component of oxide glass materials, it is considered a contaminant in chalcogenide glasses and great care is taken to remove oxygen from the host material. While some amount of oxygen will be adsorbed on the surface of the preform, it is effectively removed as the thermal energy introduced during preheat dissociates oxygen from the preform surface and is removed during the oxygen purge stage prior to pressing.

Chalcogenides consist of one or more chalcogen elements (sulfur, selenium, or tellurium) combined with a semi metallic element such as germanium, gallium, arsenic, or antimony. Selenium is a component for each of the five main types of chalcogenide glasses (ChG), although sulfur can be substituted to modify some physical and optical properties. Most suppliers of ChG use a distillation process that removes oxygen and other contaminants to purify the materials prior to melting.

Several options exist for tooling substrate materials since the process temperatures for ChG are generally lower than for oxide glasses. Materials such as Stavax, high phosphor nickel, rapidly solidified aluminum (RSA), silicon (Si), and germanium (Ge) are easily machined to create highly accurate surfaces that form intricate shapes in the molded articles. Carbide materials can still be used, but the high cost of machining and difficulty forming sharp internal transitions and intricate shapes may push molders to materials that can be diamond turned (Table 4.5).

For these types of tooling materials, protective coatings such as diamond like carbon (DLC), titanium-nitride, and titanium-aluminum-nitride coatings have shown promising results with good adhesion to the substrate over numerous molding cycles. While the titanium based coatings are generally not acceptable for oxide glass molding, they may be useful for infrared molding. The longer wavelengths associated with infrared applications are generally not as sensitive to surface and figure errors as in visible applications and may allow for looser tolerances while still yielding

Table 4.5 Common protective coatings for various substrate materials

Glass family	Substrate material	Materials commonly used as protective coatings
ChG	Carbides	Graphite, DLC, VC, Pt, Pt alloys, Re–Ir, Gold
	VC	May be used without protective coating
	Ge	DLC, VC, TiN, TiAlN, BN
	Ni	DLC, VC, TiN, TiAlN, BN
	RSA	DLC, VC, TiN, TiAlN, BN
	Silicon	SiN, Nitride based alloys, Graphite, DLC, VC, Pt, Pt alloys
	Stavax	TiN, TiAlN, BN

good performance. Final selection of the protective coating material and the application process will depend on substrate material, the glass types being produced, and customer application.

4.3 Release Coatings

4.3.1 Release Coatings—General Information

A partial list of necessary attributes for release coatings includes:

- Good adhesion to the underlying material
- Resist interaction with the heated glass (oxidation, leeching, etc.)
- Low friction between surface and heated glass
- Low wear from sliding forces
- Sufficient hardness to resist scratches
- Good resistance to both compressive and shear forces
- Good resistance to degradation due to thermal cycling.

Release coatings for molded glass articles are typically applied to the mold tool surface; however, in certain cases they may be applied to the glass preform, or to both the molding surface and the preform. The properties of release coatings may be similar to those of protective coatings, but this is not always true. For example, graphite may be a good choice for both protecting the molding tool surface and as a release agent for many oxide glasses, but glasses having a high lead content have been known to interact with the graphite and actually leech lead to the glass surface during molding to create a lead rich region that may not be acceptable for some applications. The fact that tool coatings are a necessity has been shown in literature for both inert and vacuum environments, and careful control of the molding environment will provide an effective means for preventing the oxidation of surfaces and suppressing volatile elements from diffusing from the glass and onto the mold tool surface [2].

Coatings should not appreciably increase surface roughness of the molding surface or else mechanical adhesion will become a problem. As we discuss mechanisms that promote the adhesion and release of glass to substrates and surfaces, we assume that surfaces are free from micron or larger scale surface defects that provide a mechanical means of adhesion. Surfaces must also be produced with optical quality smoothness, since rougher surfaces will enable the softened glass to gain a mechanical grab in any apparent or latent defects. When molding processes are designed for low glass viscosity at pressing, the molding pressure forces glass into voids and surface asperities, which promotes adhesion, often with catastrophic results. This condition can be ameliorated by pressing the glass at a lower temperature (higher viscosity), but this is not always possible and may in some cases lead to premature tool failure. For best results, surfaces should be polished with a maximum roughness of 2 nm RMS prior to coating.

One predictor of surface reactivity is a measure of their surface energy. This property is often determined by placing a small amount of liquid (water is commonly used) on the surface and measuring the contact angle of the droplet. The liquid is said to "wet" the surface interface when low contact angles are measured, and wetting is evidence of low surface tension between materials, low interfacial stress, and low adhesion. High energy surfaces normally form low energy interfaces with water, while low energy surfaces may be evidenced by high energy interfaces with water.

Caution should be exercised when characterizing molding surfaces with surface tension measurements, especially when they are made at temperatures much lower than those experienced during molding. Measuring surface energy with water does not guarantee that the test surface will respond in similar manner with heated glass. Some studies have shown that contact angle between glass and coating increases as temperature of the system is increased and thermal energy is added to the system, but results are highly dependent on the substrate material [2]. This outcome is surprising since the coating should be the primary interface with the heated glass, but it suggests that the chemically stable ceramic substrates also provide protection to the interface by preventing some diffusion of unwanted elements into the coating.

4.3.2 Release Coatings for Oxide Glass Applications

Most noble metals work well as release coatings for oxide based glasses, although their high cost may prove incompatible with some project goals. Platinum and platinum based alloys are the preferred materials for long life and high volume applications. Carbon in most all of its forms works well as a release coating for oxide materials as graphite, diamond like carbon, and vitreous carbon have all demonstrated good results. Coating lifetime will vary greatly with carbon based coatings due to fundamental differences in their structure and application processes. Carbon based materials, and especially carbon graphite coatings, cannot be used in oxygen rich environments since they readily join with oxygen and are thereby reduced. To ensure minimal reduction when using graphite coatings, residual oxygen levels of around 5 ppm are commonly measured in the molding chamber before heating (Table 4.6).

Application of carbon graphite coatings cost much less than platinum based coatings; however, they are a less permanent solution and may require reapplication every 300–500 cycles. The elevated temperatures encountered in molding processes serve to add energy to every component and material found therein, which increases the probability of a reaction between the carbon coating and oxygen in the glass and surrounding atmosphere. Since layers of the graphite crystal are held together by weaker forces, they tend to break down faster than the bonds found in oxide glasses. While this transition may occur slowly, it nonetheless results in a degradation of the carbon coating over time.

Table 4.6 Coatings and substrates for various oxide glasses

Glass family	Substrate material	Release coating	Comments
Oxides	WC, SiC	Pt	Long life, good release, chemically inert, expensive
		Pt alloys	Long life, high hardness, good release, chemically inert, very expensive
		Re/Ir	Long life, high hardness, good release, chemically inert, very expensive
		Gold	Average life, soft, good release, chemically inert
		VC	Average life, good release, used as substrate or coating
		DLC	Long life, high hardness, good release, chemically inert, expensive
		Graphite	Short life, soft, good release, inert w/most oxides, low cost

4.3.3 Release Coatings for Chalcogenide Glass Applications

Many additional options exist for substrates and coatings when working with chalcogenide materials. Differences in basic composition from oxide glasses and much lower molding temperatures allow for many tool processing options. Single point diamond turning (SPDT), photolithography, reactive ion etching (RIE), and other high precision processes may now be options for substrate fabrication, which increase accuracy of features and support the creation of structures such as kinoforms, freeforms, and arrays.

Materials that require grinding and polishing to create optical quality surfaces are limited in the type and size of features that can be created due to restrictions on the size of grinding and polishing tools. While CNC equipment provides increased accuracy and precision over traditional methods of manufacture, certain manufacturing rules still apply that limit feature size, sharpness of transition, minimum radius of curvature, and surface quality. For instance, the minimum diameter of grinding tools are limited by the shank diameter. As shank diameters decrease, so does their rigidity and ability to produce smooth, chatter free surfaces, which in turn places limits on substrate transition radii and surface quality. While these types of limitations are not completely removed with diamond turning processes, they are minimized to a high degree.

Manufacturing processes for germanium are well established due to extensive research through the semiconductor and infrared optics industries. Tools can be produced by several methods, and accurate features can be generated having low surface roughness and defects.

When using nickel as a substrate, high phosphor nickel (min. 14% phosphor) is recommended due to the excellent surfaces achieved from SPDT processing. Electrolytic nickel is preferred since it produces a much denser structure than electroless nickel that is deposited with more of a laminar process. Nickel is actually applied as a plated material instead of a coating, and is grown to a thickness of between 0.5 and 1.0 mm, after which it is diamond turned to produce the desired surface. The process of electrolytic deposition is not of the accuracy needed to be a standalone thin film coating, and must be machined after deposition.

RSA aluminum is usually diamond turned to produce excellent quality surfaces. Diamond turned surfaces have been demonstrated with RMS roughness of 3–4 nm, with 4–5 nm RMS roughness after nitride or DLC coatings. Aluminum oxide coatings are a natural choice since this is a surface conversion process (aluminum is converted to Al_2O_3) rather than a coating that relies on chemical and mechanical adhesion. Aluminum nitride coatings offer high hardness as well as high thermal conductivity. However, aluminum substrates may show signs of deformation and premature failure if used for extended periods at high pressing forces or in combination with process temperatures that exceed 380 °C.

Silicon tooling has benefitted from the same semiconductor industry research as germanium tools, but may have more options for coating. Silicon nitride is the most thermodynamically stable of the nitride compounds and possesses both high hardness and strength, along with excellent thermal shock resistance. Silicon is easily polished with standard optical finishing techniques, but SPDT processing may be difficult due to rapid wear of the diamond tools.

Creating optical surfaces on vitreous carbon follows similar concerns as with silicon. It responds well to traditional polishing techniques, but results have been mixed using SPDT processing. The advantage to using VC as a substrate is that it does not usually require application of a separate release coating (Table 4.7).

Release coatings are still required for most of these substrates, with coating options and restrictions both being increased with this palette of materials. Some of these substrates readily accept the coating materials; in other cases, an adhesion layer must be added; and in some cases, traditional release coatings are incompatible with the substrate.

Platinum based coatings have not demonstrated good results when molding chalcogenide materials, possibly due to interactions between platinum and selenium at molding temperatures. At elevated temperatures, thermal energy causes a change in the chemical behavior of platinum and allows it to act as a catalyst as two of the d bands are able to cross the Fermi energy level, where their chemical behavior is altered. As the Pt cools to room temperature, it returns to its original state unchanged, but the chalcogenide material undergoes a permanent chemical change.

Table 4.7 Coatings and substrates for various chalcogenide glasses

Glass Family	Substrate material	Release coating	Comments
Chalcogenide materials	VC	None	May be used without a release coating
	Ge	DLC	Long life, high hardness, good release, chemically inert, expensive. Can be applied with PVD or CVD
		VC	Average life, high hardness, good release, chemically inert. Application of coating may be difficult
		Nitrides	Long life, high hardness, good release, chemically inert
	Ni (high Phosphor)	DLC	Long life, high hardness, good release, chemically inert, expensive. Can be applied with PVD or CVD
		VC	Average life, high hardness, good release, chemically inert. Application of coating may be difficult
		Nitrides	Long life, high hardness, good release, chemically inert. Separate adhesion layer may be required
	RSA	DLC	Long life, high hardness, good release, chemically inert, expensive. Can be applied with PVD or CVD
		VC	Average life, high hardness, good release, chemically inert. Application of coating may be difficult
		Oxides	Average life, high hardness, good release, easily formed on substrate. Caution; may interact with some ChG materials
		Nitrides	Long life, high hardness, good release, chemically inert

(continued)

Table 4.7 (continued)

Glass Family	Substrate material	Release coating	Comments
	Silicon	SiN	Long life, high hardness, good release, chemically inert
		Nitride alloys	Long life, high hardness, good release, chemically inert
		Graphite	Short life, soft surface, good release, low cost
		DLC	Long life, high hardness, good release, chemically inert, expensive. Can be applied with PVD or CVD
		VC	Average life, high hardness, good release, chemically inert

4.4 Coated Preforms

4.4.1 Coated Preforms—General Information

A partial list of necessary attributes for release coatings on preforms may include:

- Very thin, perhaps <30 nm
- Easily removed
- Not appreciably increase surface roughness
- Non-reactive with glass during pressing.

Guidelines for proper selection of preform geometry are correct for both oxide and chalcogenide glass and have been previously shown (Chap. 1, Sect. 1.9). Coated preforms are not common in glass molding, but they can have a positive effect in some cases. The most common materials used as release coatings on preforms are carbon graphite for oxide glasses and alumina (Al_2O_3) for chalcogenide materials.

Information has been presented on the material properties of carbon graphite, but in order to meet the requirements listed above, different deposition processes must be considered. Pyrolytic carbon films meet these conditions since they are deposited via chemical vapor deposition techniques using gaseous hydrocarbon pyrolysis to produce an extremely thin and uniform film over the entire preform surface [3]. Most processes of this type use methane as the carbon precursor, where the film thickness is controlled by the concentration of methane gas. Removal of the carbon film is relatively simple using an oxygen plasma.

Carbon has been shown to be an effective release agent when molding oxide glasses. The main caution when applying pyrolytic carbon to the preform is that the concentration of carbon on the surface be as low as needed to provide release and that excess coating is not applied. If the coating is too thick and the displacement of glass during pressing is large, it may be difficult to remove all of the carbon from the surface of the finished lens.

The use of alumina as a release agent for chalcogenide glass has been demonstrated in published literature [4]. Results have shown that alumina is an effective release agent and provides a barrier when carbide tooling and Pt/Ir tool coatings are used, although the alumina should be effective with most tool coatings. Thickness control of the alumina appears to be more critical than the carbon-oxide glass condition, and removal of the alumina is not shown. However, the benefit of ensuring proper release between glass and tool may outweigh the detriment of minor transmission loss for certain applications.

4.5 Conclusion

The necessity of tool coatings has been demonstrated in literature and in practice, with the technology to deposit these materials as thin film coatings being well established and available in most optical shops. A wide range of substrate and coating materials is available to accommodate project requirements for quantity, accuracy, and cost. Since the composition of glasses varies greatly depending on application, careful consideration should be given regarding possible interactions at the molding interface.

References

1. E. Preston, Platinum in the glass industry. Platin. Met. Rev. **4**(1), 2–9 (1960)
2. M.K. Jeng et al., Design of protective coatings for glass lens molding. Key Eng. Mater. **364–366**, 655–661 (2008)
3. T. Kaplas and P. Kuzhir, Ultra-thin pyrocarbon films as a versatile coating material. Nanoscale Res. Lett. **12**, 121 (2017)
4. H.H. Chien et al., Molding of Al_2O_3 coated chalcogenide glass lenses. Opt. Eng. **51**(3), 033401 (2012)
5. S. Ployhar et al., ITER Components Cooling: satisfying the distinct needs of systems and components, Fusion Eng. Des. (2014)

Chapter 5
Moldable Glasses

The development of glass formulations made specifically for molding applications, combined with the introduction of commercially available molding equipment, has made production molding an economical and viable process for nearly all manufacturers of precision optics.

5.1 Background

The formation of optical quality surfaces from a softened piece of glass may have been understood toward the end of the nineteenth century, but was of little practical value due to the limited availability of optical glass types to impact potential applications of optical systems. But in 1884, optical industry pioneers Otto Schott, Ernst Abbe, and Carl Zeiss formed the "Schott and Associates Glass Technology Laboratory" that began developing glasses with specific optical properties that had both stable and reliable means for production. Their groundbreaking work of transforming optical glass into a technical material stimulated the need for commercial manufacturing of precision optical surfaces.

While glass molding of precision optics has been a viable production process for over 100 years, the development of glass molding processes closely followed the work of Otto Schott and others who provided optical engineers with a path to realize their designs. What began as a novel invention with few applications has developed into a science based industry with the ability to transform optics manufacturing.

Like many industries, glass molding continues its reliance on supporting technologies to fully realize its potential. Precision machining of tool materials, thin film coating development, and precision machine actuation are all basic requirements for consistently producing accurate surfaces. These technologies and many others have developed through the years and enabled glass molding to reach its present state.

J. J. Nelson, *Precision Lens Molding of Glass: A Process Perspective*,
Progress in Optical Science and Photonics 8,
https://doi.org/10.1007/978-981-15-4238-1_5

As industry continues to improve on previous discovery and reveal new applications for molded optics, development of glasses that are amenable to molding has broadened the market space for potential manufacturers. The availability of these moldable glasses has supported manufacturing by reducing R&D investment with materials having lower process temperatures and carefully chosen properties that empower both manufacturing and design with glasses that encompass an ever widening range of the Abbe diagram.

Moldable glasses did not exist much before the 1990s, yet companies were able to produce high quality optics with "standard" glass types. How was this possible, and what advantages do moldable glasses bring to the industry? What makes one glass more favorable to molding than another? What makes a glass "moldable", or suitable for use in glass molding processes? What characteristics do "moldable glasses" possess that differ from "non-moldable glasses"? Are non-moldable glasses truly not able to be molded into optics? Answers to these questions will be the focus of this chapter.

5.2 Chemical Compositions

5.2.1 Oxide Glasses

For the purposes of this section, we focus on optical glasses that have application mainly in the visible portion of the electromagnetic spectrum, have no noticeable ceramic composition, and whose principal optical properties are index of refraction, dispersion, and transmission. Elements that combine with oxygen to act as glass formers are boron, silicon, phosphorus, germanium, arsenic, and antimony. Elements such as aluminum, titanium, vanadium, gallium, selenium, molybdenum, tellurium, tungsten, and bismuth all act as intermediates, or conditional glass formers. These oxides do not form glasses by themselves, but act like glass formers when combined with other elements (i.e., aluminosilicates) (Table 5.1).

Optical properties, thermal properties, physical properties, and chemical resistance of the glass can all be adjusted by making slight changes in composition. For

Table 5.1 Periodic table of the elements with oxide glass forming elements noted

IA	IIA	IIIB	IVB	VB	VIB	VIIB	VIII			IB	IIB	IIIA	IVA	VA	VIA	VIIA	0
H																	He
Li	Be											B	C	N	O	F	Ne
Na	Mg											Al	Si	P	S	Cl	Ar
K	Ca	Sc	Ti	V	Cr	Mn	Fe	Co	N	Cu	Zn	Ga	Ge	As	Se	Br	Kr
Rb	Sr	Y	Zr	Nb	Mo	Tc	Ru	Rh	Pd	Ag	Cd	In	Sn	Sb	Te	I	Xe
Cs	Ba	La	Hf	Ta	W	Re	Os	Ir	Pt	Au	Hg	Tl	Pb	Bi	Po	At	Rn

Table 5.2 Periodic table with general trends for oxide glasses

Large Bandgap → High v

Higher Density → High n

IA	IIA	IIIB	IVB	VB						IIB	IIIA	IVA	VA	VIA	VIIA	0
H																He
Li	Be										B	C	N	O	F	Ne
Na	Mg										Al	Si	P	S	Cl	Ar
K	Ca	Sc	Ti	V	Cr	Mn	Fe	N	Cu	Zn	Ga	Ge	As	Se	Br	Kr
Rb	Sr	Y	Zr	Nb	Mo	Tc	Ru	Pd	Ag	Cd	In	Sn	Sb	Te	I	Xe
Cs	Ba	La	Hf	Ta	W	Re	Os	Pt	Au	Hg	Tl	Pb	Bi	Po	At	Rn

instance, the refractive index of glass tends to track with density, so glasses that contain greater amounts of heavier elements possess higher density, and generally have a higher refractive index. This makes glass an attractive choice over materials such as crystals whose properties are for the most part fixed by their crystallographic structure (Table 5.2).

The building block for most oxide glass is silica (SiO_2), either by itself or with an intermediate glass former. Information on the molecular scale structure of silicate glasses is debated even today, as no agreement exists to its precise nature and principles of formation since they possess no long range consistency in structure or form. Instead, we choose to examine short range order for all glass types in an effort to understand their dynamic and thermodynamic properties.

Aluminosilicate glasses possess varying amounts of alkali or alkali earth elements such as MgO or CaO, often between 5 and 15%. These glasses are characterized by high elastic moduli and resistance to chemical attack, with electrical resistance equal to or greater than that found in vitreous silica.

Borosilicate glasses are formed with relatively small amounts of alkali elements and are known for having a very low coefficient of thermal expansion and high resistance to chemical attack. Lead silicate glass is characterized by high refractive index, high dispersion, and large working range. This extended working range makes molding process development easier since their viscosity responds slowly to changes in temperature.

Phosphate glasses have P_2O_5 as a major compound and are useful for many biomedical and optical applications including components for high energy laser systems. They are also resistant to attack by hydrofluoric acid, a property that is not common among oxide glasses. Their main disadvantage is their overall poor chemical durability, a propensity to devitrify when heated above the transformation temperature, and the tendency to be corrosive to some refractory linings, resulting in challenges during production. In regard to grinding and polishing, phosphate glasses exhibit different behavior than silicate materials and are processed differently than most oxide glasses. Since glasses containing a high percentage of P_2O_5 often possess substantial amounts of water, small amounts of fluoride are sometimes added to drive off water during the batch melting process.

Fluorosilicate glasses find many applications as optical fibers, especially when used at wavelengths above 1 μm. The presence of fluoride lowers the optical dispersion of the glass and frequently reduces the transformation temperature of the melt.

Lanthanum based compositions are characterized by high refractive index, low dispersion, and good transmission in the shorter wavelengths and are typically used in higher end optical systems. Some strengths of titania glasses are enhanced chemical durability, low dispersion, and refractive index values that can be greater than 2.0. Many more glass types exist as glass scientists continue to look at means of expanding current offerings for new applications and requirements.

Just as it is obvious that composition affects intrinsic properties of the melt, thermal properties such as the transformation temperature (T_g, 10^{12} Pa s) and Littleton softening point (T_s, $10^{6.6}$ Pa s) can be modified by slight changes in stoichiometry. While boric oxide (B_2O_3) increases thermal stability and chemical durability, it also causes an increase in transformation temperature as more energy is required to loosen the interatomic bonds. The presence of alumina (Al_2O_3) will cause a marked increase in transformation temperature and is normally found in lower concentrations for moldable glass types (Table 5.3).

Some compounds commonly found in low temperature oxide glasses are phosphorus pentoxide (P_2O_5), alkali metal oxides, or alkaline earth metal oxides. Alkali metal oxide concentrations are commonly held between 15 and 45%, with alkaline earth metal oxide concentrations of 10% or less. In melts where P_2O_5 is less than 30%, B_2O_3 may be incorporated up to 25%, but will cause an increase in T_g and T_s properties. Optional oxides such as lead oxide (less than 20%), lanthanum oxide (7% or less), and zinc oxide (12% or less) may be found, with the total of all optional ingredients being less than 35% (Table 5.4).

In recent years, glass manufacturers have focused efforts on developing formulations specifically designed for precision glass molding processes while expanding the available range of materials on the Abbe diagram. These glasses typically have transformation temperatures below 550 °C with compositions designed to be both

Table 5.3 Exemplar oxide bond strength for various glass forming compounds

	Bond strength (kJ/mol)		Bond strength (kJ/mol)
B_2O_3	−1254	MgO	−601
SiO_2	−911	CaO	−635
Al_2O_3	−1675	SrO	−592
P_2O_5	−1527	BaO	−582
Li_2O	−595	PbO	−202
Na_2O	−416	La_2O_3	−1791
K_2O	−363	ZnO	−350

Table 5.4 Exemplar stoichiometry of moldable oxide glasses

				Alkali metal oxides			Alkali Earth metal oxides				
	B_2O_3	SiO_2	P_2O_5	LiO	Na_2O	K_2O	MgO	CaO	SrO	BaO	F
Max%	–	–	50	45	20	10	10	15	20	20	3

environmentally friendly and less reactive with standard tool materials [1]. They are also offered in shapes and sizes commonly sought after by molding companies.

Boron based compositions are characterized by lower density (boron has lower atomic mass than other glass formers on the periodic table) and are popular for their stability over rapid changes in temperature, which is due to their relatively low coefficient of thermal expansion and high heat capacity. However, a slight change in composition can alter the softening point of the material by almost 100 °C, making it attractive for molding applications (*note P-BK7 and N-BK10 in* Table 5.5).

Many glass manufacturers that have developed product lines specifically for glass molding operations designate them by special codes. For example, SCHOTT uses the "P" prefix to denote glasses designed for molding, O'Hara uses the prefix "L", and Hoya uses the prefix "M". In addition to having low transformation temperatures and softening points within the range of most commercial equipment, these glasses are provided in forms suitable for molding, most often as ball preforms or gobs (Table 5.6).

One advantage of ball preforms is that no orientation is required when placed in a concave mold since the ball will naturally find the vertex of the concave surface. Gobs do not have a singular curvature and will generally have one surface with a controlled radius while the other is allowed to form under surface tension. Gobs will not find the vertex of a concave surface and must be aligned and oriented in the mold tooling to ensure symmetric flow and good results. Glass volume is well controlled in both processes. In addition to having favorable geometries, both of these preforms are normally lower cost than ground and polished types since they are manufactured in higher volume and with bulk processing techniques (Fig. 5.1).

In addition to glasses made especially for molding applications, many manufacturers offer glasses with transformation and molding temperatures that fall in the process window of commercial equipment, but are not offered in preform shapes that are amenable to molding. Examples of these would be SCHOTT N-FK51A, N-PK52A, N-PK51, and N-KZFS2. These glasses have physical and thermal properties that compliment molding equipment and may still be processed into finished lenses, but preforms must be ground and polished from bulk material, often at significant cost compared with high volume preform manufacture.

Because preform specifications may actually be similar to those of finished lenses, preform selection can be an important factor to meeting overall production goals. Making use of shapes that are supported by well-developed manufacturing processes will ensure the lowest production costs for the finished optic (ref Fig. 1-8).

Characteristics most impactful for molding are thermal properties, followed by physical and chemical properties. Glasses with high molding temperatures and high

Table 5.5 Properties of various borosilicate glass types

	Optical properties		Thermal properties			Physical properties			Chemical properties				
	n_d	v_d	T_g (°C)	$T_{10^{7.6}}$ (°C)	α (10^{-6}/°C)	ρ (g/cm³)	C_p (J/g K)	K (W/m K)	CR	FR	SR	AR	PR
N-BK7	1.5168	64.17	557	719	8.3	2.51	0.858	1.14	1	0	1	2.3	2.3
P-BK7	1.5164	64.06	498	657	7.3	2.43	0.87	1.13	1	0	1	2.3	2.3
N-BK10	1.49782	66.95	557	753	6.6	2.39	0.81	1.32	1	0	1	1	1

Table 5.6 Exemplar selection of moldable glass types

SCHOTT	Glass code	O'Hara	Glass code	SCHOTT	Glass code	O'Hara	Glass code
P-BK7	516641	L-BSL7	516641			L-LAM69	731405
P-PK53	527662					L-LAM60	743493
		L-BAL42	583594	P-LAF37	755457		
P-SK57Q1	586595	L-BAL42P	586592			L-LAH91	765491
		L-BAL43	586597	P-LASF47	806409	L-LAH53	806409
P-SK57	587596					L-LAH84	808405
P-SK58A	589612	L-BAL35	589612	P-LASF50	809405		
		L-BAL35P	592610	P-LASF51	810409		
P-SK60	610579					L-LAH90	832401
P-SF8	689313	L-TIM28	689311			L-LAH85V	854404
P-LAK35	693532	L-LAL15	693529			L-LAH94	861371
		L-LAL13	694532			L-LAH86	903310
		L-TIM28P	695307	P-SF67	907214		
P-SF69	723292			P-SF68	005210		

Fig. 5.1 Oxide glass balls and spherical chalcogenide preforms. *Courtesy* SCHOTT AG

densities generally require longer cycle times and often necessitate tool materials that can withstand high thermal cycling. This may in turn limit the type of features that can be cut into the tool. Glasses with extreme thermal expansion properties may need to be matched with tools having similar characteristics.

5.2.2 Government Regulations Affecting Glass Manufacture [2, 3]

Two regulations having the most impact on glass production are known as REACH (EC 1907/2006 of 18 December 2006) and RoHS (2002/95/EC of 27 January 2003).

Table 5.7 Glasses currently exempted from RoHS directive

Optical glasses affected by RoHS directive		
K7	LAFN7	SF5
K10	KZFSN5	SF6
LLF1	KZFS12	SF10
LF5	SF1	SF56A
F2	SF2	SF57
F5	SF4	

REACH (Regulation on Registration, Evaluation, Authorization, and restriction of Chemicals) is an effort to improve the protection of human health and environment through early identification of intrinsic properties of chemical substances, and places responsibility on industry to manage risks from chemicals. If REACH were to be fully enforced, only five glass types in the SCHOTT catalog would be permitted of the over 120 currently available. The impact of this directive is obvious for the optics industry.

The chemicals identified in the directive that most impact glass manufacture are arsenic oxide, lead oxide, and boron oxide. Glass makers have found substitutes for these chemicals in some applications, but not all. The impact on product offerings extends to nearly all optical glasses, including low T_g glasses.

RoHS (Removal of Hazardous Substances) seeks to restrict the use of hazardous substances in electronics and electronic equipment, and promotes recycling in an effort to reduce increasing amounts of these chemicals in landfills. Since many optical systems are mated with electronics, the directive extends to glass materials as well. The glasses affected by this directive are those which contain lead, arsenic, antimony, or compounds of these elements, so RoHS encompasses a much smaller range of glass types than the REACH directive.

While this directive did not affect the current offering of moldable glass types, its puts restrictions on the development of future low T_g and moldable glasses. Table 5.7 lists glass types that were deemed to be indispensable and have been temporarily exempted from the directive.

Both of these directives have been periodically reviewed and revised after assessing the impacts of original timelines. It is understood in the glass making industry that any potential hazards posed by these elements are extremely minimal. While these elements may be dangerous at some level, they become completely bound into the glass matrix and are frozen at the atomic level, unable to combine or react with anything outside of that matrix. As a result, some extensions and exemptions have been granted, although not indefinitely.

5.2.3 Compatibility with Coatings

Most of the tool coatings previously described will perform well with the majority of moldable glass types. However, there are some compounds that merit consideration.

It was previously noted that P_2O_5 may be a component of some moldable glasses, but this compound has demonstrated low level reactivity with some refractory materials (i.e., some noble metals). Therefore, premature surface staining and pitting may occur when using certain tool coatings with moldable glass types having a high concentration of P_2O_5. The amount of apparent tool surface degradation will depend on the concentration of available phosphorus oxide, but this reaction should be considered when molding these materials. If noble metals are used for tool coatings and the tool surface begins to worsen, then the use of DLC coatings may prove to be a better choice for that particular application.

5.2.4 Infrared Glasses

When we speak of glasses made for infrared (IR) applications, we refer to the family of chalcogenide glasses named for the chalcogen elements oxygen, sulfur, selenium, and tellurium. Oxygen is not included as a component in chalcogen glasses since many oxygen compounds are considered impurities and absorbers at infrared wavelengths (see Fig. 5.2).

Chalcogenide glass is made by combining one or more of the chalcogen elements with a semi-metallic element such as gallium, germanium, arsenic, or antimony. The bonds formed by these elements are mostly covalent in nature, with weak van der Waals attraction between molecules (Table 5.8).

Most chalcogenide glasses are considered to be moldable due to their relatively low transformation and molding temperatures and availability of preforms for molding. However, some glass types exist that transmit in limited regions of the infrared and are not chalcogen based. Certain formulations of heavy metal oxides such as lead

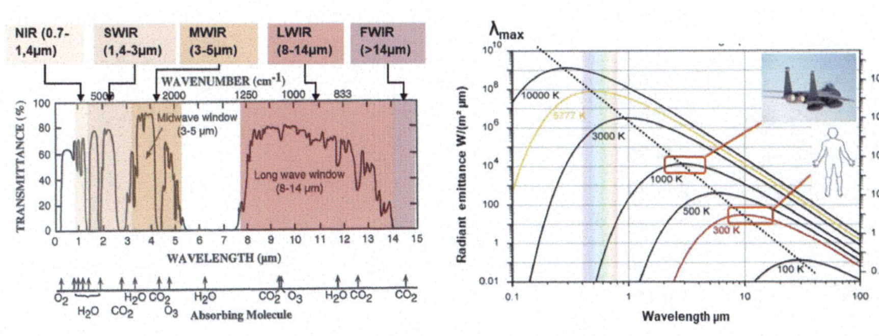

Fig. 5.2 Absorbing molecules and infrared regions. *Courtesy* SCHOTT AG

Table 5.8 Periodic table of the elements with chalcogenide glass forming elements noted

Semi Metallics

Chalcogen Elements

IA	IIA	IIIB	IVB	VB	VIB	VIIB	VIII			IB	IIB	IIIA	IVA	VA	VIA	VIIA	0
H																	He
Li	Be											B	C	N	O	F	Ne
Na	Mg											Al	Si	P	*S*	Cl	Ar
K	Ca	Sc	Ti	V	Cr	Mn	Fe	Co	N	Cu	Zn	Ga	Ge	As	*Se*	Br	Kr
Rb	Sr	Y	Zr	Nb	Mo	Tc	Ru	Rh	Pd	Ag	Cd	In	Sn	Sb	*Te*	I	Xe
Cs	Ba	La	Hf	Ta	W	Re	Os	Ir	Pt	Au	Hg	Tl	Pb	Bi	Po	At	Rn

bismuth oxide glass (Pb–Bi–O) and tellurite oxide glass (Te–O) transmit well from the near infrared to mid-infrared regions but have not found broad usage due to their limited transmissive windows and availability. These materials generally have good transmission between 550 nm and 5 μm, making them useful for NIR, SWIR, and some MWIR applications that also benefit from limited transmission in the visible (Table 5.9).

Another advantage is that they are produced with more conventional melting methods; that is, they are not constrained to the size of an ampoule and rocking station, as are many chalcogenide glasses. Therefore, they can be produced in larger formats and have a cost structure more closely related to standard oxide glass melts. They can also be doped with active laser gain elements that find application in laser systems. However, these materials are considered "specialty materials" and preforms are not offered in geometries or processes that are amenable to molding.

Heavy metal fluoride glasses are a form of non-oxide glasses with application primarily in the NIR to MWIR bands of the infrared. Their main usage has been as fibers for long haul telecommunications, with few needs for optical components such

Table 5.9 Optical properties of some common chalcogenide glasses

	Refractive index (η)	Transmission start (μm)	Transmission end (μm)
$Ge_{22}As_{12}Se_{55}$	2.4967	0.79	11.6
$Ge_{30}As_{13}Se_{32}Te_{25}$	2.7869	1.3	12.0
$Ge_{10}As_{40}Se_{50}$	2.6090	0.8	11.8
$Ge_{28}Sb_{12}Se_{60}$	2.6032	0.93	12.1
$As_{40}Se_{60}$	2.7782	0.9	13.2
Pb–Bi–O	1.8696	0.4	4.6
Te–O	1.9041	0.36	6.0

*Transmission start and end values approximate, considered at 40% level
*Refractive index measured at 10 μm for ChG; 1.5 μm for HMO glasses

as couplers and collimators to be made from the same material. They are, however, extremely prone to crystallization and attack by water. While molded articles may be produced with these materials, they are not considered moldable glasses since preforms are only available through standard grinding and polishing processes.

A common goal of infrared transmitting materials is to extend the transmission range as far as possible on both ends of the spectrum. This enables greater application of the material and in some cases lessens the need for additional materials and optical components. Additional requirements of IR glasses are:

- High transmission in the infrared
- Greatest extension of transmission through the IR
- Free from IR absorption bands
- Consistent dispersion across multiple IR spectral bands
- Good resistance to environment (moisture, environmental chemicals, etc.)
- Thermal stability through a wide temperature range
- Adequate hardness
- Low scattering loss.

Transmission ranges in ChG are determined by the energy levels of elements in the material. Energy in a glass molecule consists of three main components; electronic energy, rotational energy, and molecular energy. The absorption of a glass in the shortwave region is generally accomplished by electron energy transitions, while absorption in the longwave region is caused by inner molecule or lattice vibrations (i.e., heavier elements push it towards longer cutoff transitions). Molecular rotational energy is linked to absorption in the far infrared region and will not be discussed. In general, transition frequencies identify the start of a transmission region, while oscillations of heavier elements define the cutoff transmission.

Consistent dispersion is a difficult requirement for many optical materials due to the enormous spectral bandwidth of the IR. While the visible spectrum is considered to be less than 400 nm in length, infrared bands (780 nm to 1 mm) extend more than 2500 times the range of the visible band. In general, chalcogenide glasses are better suited to meet dispersion requirements than most IR transmitting crystals. Whereas many IR transmitting crystals possess a wavelength dependent dispersion, ChG materials have nearly constant dispersion over wide spectral range.

5.3 Desirable Properties for Precision Glass Molding Processes

5.3.1 Thermal Properties

A partial list of desirable thermal attributes for moldable glasses are:

- Long working range; i.e., the difference between molding temperature and T_g

- Relatively low molding temperature
- Good thermal conductivity
- Good thermal diffusivity.

Glasses having a long working range allow for more processing options and make process control much easier. Materials having a relatively low molding temperature allow the use of tooling materials and coatings that are not appropriate for higher temperature materials. Glasses that possess higher thermal conductivity and diffusivity generally have lower process cycle times.

Recall from Chap. 2 that volumetric thermal diffusivity κ (m^2/s), is a measure of thermal stability for non-homogeneous temperature distributions within the system. Materials with high diffusivity values are preferred since they transfer heat more rapidly, which allows them to quickly achieve equilibrium with their surroundings, and may result in less distortion of the molded article. The thermal diffusivity of oxide glasses typically range between two and five times greater than for chalcogenide materials, with common tool materials transferring thermal energy at ten to one hundred times the rate of most moldable oxide glasses (Table 5.10).

5.3.2 Physical Properties

A partial list of desirable physical attributes for moldable glasses are

- Good hardness
- Relatively high fracture toughness

Table 5.10 Thermal properties of some common oxide and chalcogenide glasses

	CTE (ppm/K)	C_p (J/g K)	K (W/m K)	κ (m^2/s)	T_g (°C)	T_M (°C)
P-BK7	7.30E−06	0.87	1.13	5.35E−04	498	657
P-PK53	1.60E−05	0.77	0.64	2.94E−04	383	453
P-SK57	8.90E−06	0.76	1.01	4.42E−04	493	593
P-LASF47	7.30E−09	0.55	0.85	3.40E−04	530	627
P-SF67	7.40E−06	0.53	0.79	3.52E−04	539	663
N-FK51A	1.48E−05	0.69	0.76	2.99E−04	464	527
N-FK5	1.00E−05	0.81	0.93	4.67E−04	466	672
$Ge_{22}As_{12}Se_{55}$	1.21E−05	0.33	0.24	1.65E−04	368	445
$Ge_{30}As_{13}Se_{32}Te_{25}$	1.34E−05	0.32	0.22	1.42E−04	275	360
$Ge_{10}As_{40}Se_{50}$	2.04E−05	0.37	0.18	1.09E−04	225	310
$Ge_{28}Sb_{12}Se_{60}$	1.40E−05	0.33	0.25	1.63E−04	285	348
$AS_{40}Se_{60}$	2.08E−05	0.36	0.24	1.44E−04	185	236
Pb–Bi–O	7.60E−06	NA	0.53	–	449	–
Te–O	1.86E−05	NA	NA	–	249	–

Table 5.11 Physical properties of some common oxide and chalcogenide glasses

	Density (g/cm^3)	H_k (10^{-6} mm^2/N)
P-BK7	2.43	627
P-PK53	2.83	335
P-SK57	3.01	535
P-LASF47	4.54	620
P-SF67	4.24	440
N-FK51A	3.68	345
N-FK5	2.45	520
$Ge_{22}As_{12}Se_{55}$	4.41	143
$Ge_{30}As_{13}Se_{32}Te_{25}$	4.84	136
$Ge_{10}As_{40}Se_{50}$	4.47	112
$Ge_{28}Sb_{12}Se_{60}$	4.66	113
$As_{40}Se_{60}$	4.63	110
Pb–Bi–O	5.88	384
Te–O	5.03	NA

Materials having superior hardness are more resistant to the formation of scratches and surface defects. High fracture toughness results in components that are less prone to breakage or chipping, either from handling or post processing operations such as centering or thin film coating (Table 5.11).

5.3.3 *Chemical Properties*

A partial list of desirable chemical attributes for moldable glasses are

- Good resistance to water
- Good resistance to environmental chemicals.

Most non-silicate glasses are likely to react with water, and those that show resistance to staining by water are often susceptible to reaction with acids or strong bases. Glasses that contain even moderate levels of alkali ions will frequently react with H_2O ions in the slurry or surrounding atmosphere and either dissolve into the neighboring environment or deposit a film, or stain, on the glass surface.

As with many processes, the ion exchange process between the glass surface and its immediate environment is diffusion limited, which means that the extent of damage into the glass surface increases as the square root of time. For silicate based glasses, pH levels above 10 and below 2 may present an increased danger of reaction and damage to the glass surface.

Glasses are tested and rated by the manufacturer and are generally consistent between suppliers, although small differences may be encountered. Resistance to

Table 5.12 Chemical properties of some common oxide glasses

	CR	FR	SR	AR	PR
P-BK7	1	0	1	2.3	2.3
P-PK53	2	1	51 (a)	4.3	4.3
P-SK57	4	3	52.3 (a)	2	3
P-LASF47	1	1	51.4 (a)	1	2.2
P-SF67	1	D	1	1.3	1
N-FK51A	1	0	52.3 (a)	2.2	4.3
N-FK5	2	1	4	2	2.3

CR = climatic resistance at high humidity and temperatures (40–50 °C)
FR = stain resistance using a mildly acidic buff er solution (pH = 4.6, 5.6)
SR = acid resistance using a highly acidic solution (pH = 0.3)
(a) = acid resistance using a moderate acidic solution (pH = 4.6)
AR = alkali resistance using a moderate alkali solution (pH = 12) at 50 °C
PR = phosphate resistance using a mildly basic solution (pH = 10) at 50 °C

water will minimize the effects of airborne moisture affecting the optical surface. Resistance to environmental chemicals will enable more choices for cleaning and will resist the formation of stains both before and after coating. Information for infrared transmitting glasses were not available (Table 5.12).

5.4 Index Drop

5.4.1 Causes of Index Drop

Index drop is a natural outcome of the molding process, regardless of whether the glass is a standard oxide glass, moldable glass, or chalcogenide glass. Glass is cooled much more quickly in the precision lens molding process than during initial manufacture of the glass. Faster cooling freezes in properties more consistent with the liquid state and has a direct effect on the refractive index of the glass.

As the tool package is heated through the molding cycle and thermal energy is added to the glass, vibrational amplitude of the molecules is increased, and the material expands at a fairly regular rate. Eventually, this energy fills the available degrees of freedom (DOF) for the glass molecules (molecules are vibrating at maximum amplitude), and additional thermal energy is accommodated by opening up additional DOF of motion. Rotational and translational DOF now become available, and the glass expands at a much accelerated rate. This transition between available degrees of freedom is where we find the glass transition temperature (T_g). The slope of this curve over a defined temperature range specifies the coefficient of thermal expansion (CTE).

The transition from glassy state to supercooled liquid is a kinetic transition that is regulated by the viscosity characteristics of the glass. Viscosity can be thought of as a measure of the internal friction between molecules that move in a direction to resist the external force, and therefore, both internal and external processes drive the system. External processes are defined by the heat source and power available to input thermal energy to the system. Internal processes are regulated by bonding at the molecular level; covalently bonded materials exhibit strong structures requiring higher energy to loosen and higher T_g characteristics, while molecular or ionic bonds are weaker and will separate with lower thermal energy input.

Another way to look at this is by considering the heat capacity of the glass at various stages. Heat capacity curves possess similar shape to the curve shown in Fig. 5.3. The heat capacity of glass is related to the vibrational contributions of the individual atoms. But heat capacities of liquids are defined by vibrational, rotational, and translational degrees of freedom. Even though the molding temperature of glasses are often in the range of the Littleton softening point (T_s, $10^{6.6}$ Pa s) and the glass is not in a purely liquid state, it begins to exhibit attributes usually associated with systems at higher temperature.

Once the glass is heated to its molding temperature and the article is pressed to its final shape, the system begins its cooling cycle. As the glass is cooled, mass stays constant but volume decreases with the decrease in temperature, which causes an increase in density. If allowed to cool slowly, refractive index of the molded glass would approach that of the bulk material. But to avoid devitrification and increase throughput, cooling rates are often accelerated.

Rapid cooling decreases the time available for the semi-liquid to adjust to the temperature change and the properties consistent with that lower temperature. The glass structure is frozen since the molecules are not able to overcome the viscous forces that restrict their continued motion. Since refractive index is directly related

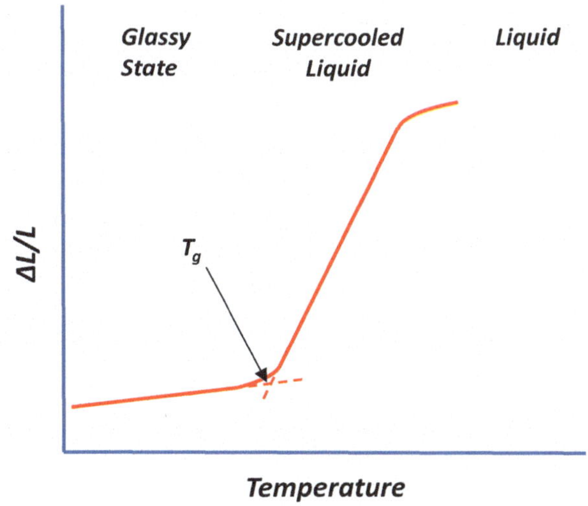

Fig. 5.3 Exemplar temperature curve for glass

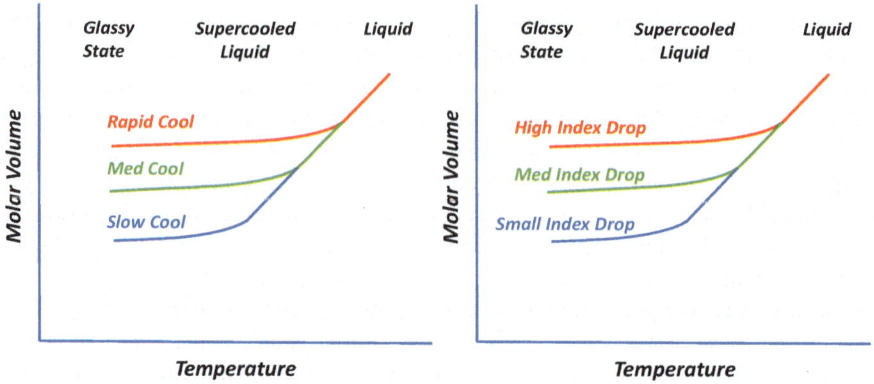

Fig. 5.4 Exemplar cooling curves for glass

to density, faster cooling leads to a higher volume of material, lower density, and lower refractive index and are responsible for higher values of index drop. As shown by Fig. 5.4, the actual index drop experienced for a particular lens geometry and material is highly dependent on the specific molding process employed.

5.4.2 *Oxide Glasses*

Index drop values for oxide glasses are generally in the range of a few parts per thousand. Actual values will vary with the composition of the glass and process parameters chosen, with special emphasis on the rate of cooling (Table 5.13).

Table 5.13 Exemplar index drop for oxide glasses

Material	T_m range (°C)	Index drop (n_d)
P-PK53	425–445	−0.00370
P-SK57	555–580	−0.00270
P-LASF47	595–610	−0.00450
P-SF67	610–630	−0.00700
N-FK51A	505–525	−0.00186
N-FK5	600–630	−0.00249
N-PK52A	515–535	−0.00180
N-PK51	540–560	−0.00185

Table 5.14 Exemplar index drop for ChG glasses

Material	T_m range (°C)	Index drop ($n_{10\mu m}$)
$Ge_{30}As_{13}Se_{32}Te_{25}$	320–340	−0.00324
$Ge_{10}As_{40}Se_{30}$	270–290	−0.00264
$Ge_{28}Sb_{12}Se_{60}$	325–345	−0.00135
$As_{40}Se_{60}$	215–235	−0.01209

5.4.3 Infrared Glasses

Index drop values for chalcogenide glasses are generally in the range of a few parts per thousand, although some materials may be significantly greater. In particular, compositions of As–Se and As–S will exhibit much higher changes in refractive index due to their internal bonding structures and nature of their compositions. As with oxide glasses, actual values will vary with the composition of the glass and process parameters chosen (Table 5.14).

5.4.4 Index Homogeneity

Discussions regarding material homogeneity need to distinguish between long range spatial homogeneity and short range deviations in homogeneity such as striae. The introduction of striae other than what is present in the preform should be non-existent from the molding process as long as the flow and movement of the glass during pressing is well controlled. If the glass is allowed to "roll over" during pressing, then striae may be evidenced as a result of poor process control.

Long range spatial homogeneity of refractive index in articles produced by the glass molding process should be mostly consistent with that of the bulk material. The main causes of long range homogeneity variations are

- Gradients in chemical composition produced during melting
- Variations in density due to thermal equilibrium and thermal history
- Permanent stress due to temperature gradients during cooling

Of these three, molding may have the greatest effect on the creation of permanent internal stress. But this should be controlled during the molding process, and since the glass is never melted and only softened before press out, it is unlikely that these thermal stresses would be a major concern. It is conceivable that stress could be formed by pressing out the lens too quickly and building up excessive internal stress in the material, but once again, this is controlled by creating reasonable molding processes that consider the nature and sensitivities of the material.

As discussed later in Sect. 7.1, the molding process creates a naturally tempered article since the surface of the glass is first to be frozen while the interior continues to cool. This difference in cooling rate will cause a difference in the refractive index

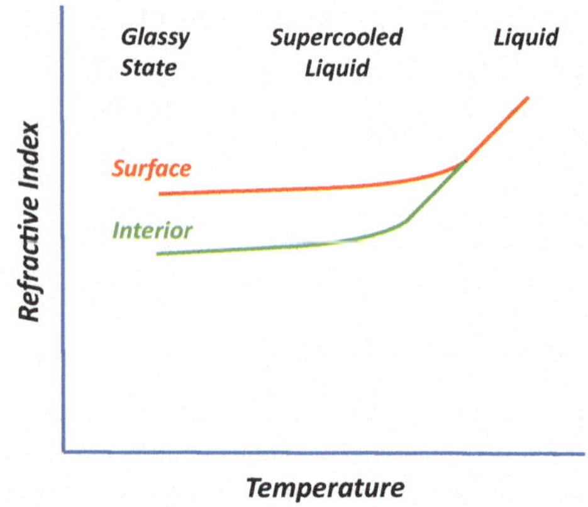

Fig. 5.5 Refractive index variation within a molded article

from the center to surface of the glass article. In most cases, this is not a concern as a noticeable change of index may only appear in the 4th or 5th decimal place. However, this effect should be considered when working with extremely demanding applications having high sensitivity to variations of refractive index (Fig. 5.5).

5.5 Special Handling

Once the desired surfaces have been molded into the glass, the finished article should be treated carefully, as should any precision optic. Excessive handling will likely lead to the formation of scratches and other unwanted artifacts. In general, oxide glasses have greater surface hardness than chalcogenide materials, so extra care should be given when working with infrared transmitting materials.

Surfaces of the molded lenses will never be cleaner than when they are removed from the mold, except in cases where release coatings have been applied to the preform. Otherwise, the lens should be free from surface contaminants. The molded articles should be inspected for surface defects and then packaged for transport or prepared for application of thin film coatings.

Cleaning agents commonly used for oxide glasses should work well with moldable oxide glasses. Hand cleaning and ultrasonic cleaning are both common methods that are used throughout the industry. However, special care should be exercised when working with chalcogenide materials.

Whereas oxide glasses are generally reactive to acids and resistant to bases, chalcogenide materials function in the opposite manner. Also, due to the low thermal conductivity and high thermal expansion of ChG materials, the use of acetone as a cleaning agent should be avoided when working with these materials. Acetone evaporates much faster than IPA, which results in a rapid change in temperature at the surface that could damage the optic.

Standard Oxide Glass

- Generally harder materials ($H_k \approx 600$)
- Low to moderate thermal expansion
- Moderate thermal conductivity
- Low to moderate fracture toughness
- Reactive to acids
- Resistant to bases
- Lower refractive index
- Low thermal optic coefficient
- Wavelength dependent dispersion.

Standard Chalcogenide Glass

- Generally softer materials ($H_k \approx 110$)
- High thermal expansion
- Low thermal conductivity
- Low fracture toughness
- Resistant to acids
- Reactive to bases
- Higher refractive index
- Moderate to high dn/dT
- Constant dispersion over wide range.

5.6 Conclusion

The last 30 years have seen a large increase in moldable glass types produced by glass manufacturers for both visible and infrared applications. Commercial equipment has also entered the marketplace to make glass molding a potential capability for many producers of high quality optics. As developments in tooling materials and thin film coatings continue to expand, it is likely that additional offerings will become available for suppliers of molded glass optics.

References

1. SCHOTT TIE 40, *Optical Glass for Precision Molding*, V6-2, SCHOTT AG, Sept 2011
2. P. Hartmann et al., *REACH and RoHS May Endanger Availability of Raw Materials for Glasses*, 01 Apr 2014
3. P. Hartmann, U. Hamm, *Optical Glass and the EU Directive RoHS*, 20 Apr 2011. https://doi.org/10.1117/12.882922

Chapter 6
Crystallization Kinetics

Formation of the glassy solid is not a normal thermodynamic process since the elements frozen in this form are not at their lowest and most preferred energy state. Glass is only formed as we deviate from thermodynamic stability, and the elements will readily return to a crystalline structure if they are allowed to rearrange to a low energy state.

All glasses as we have defined them are inherently thermodynamically unstable. Their molecules have been forced into an unstable state that cannot be overcome due to the high viscosity, or internal friction, that limits their movement (see Sect. 5.4.1). While crystallization of the glass upon cooling is not usually of great concern, it should be noted that this effect does not need to occur throughout the entire volume to be problematic. Even small areas of crystal growth can negatively affect the optical and physical properties of the finished lens.

Small regions of crystal formation can be evidenced as scattering sites that impact transmission characteristics, while larger areas of crystal growth will affect physical and thermal properties such as thermal expansion and birefringence. Crystals generally possess a smaller volume than their glassy counterparts, which will affect density and refractive index properties. Crystal regions within the glass solid may be seen as imperfections in the melt, striae, or some other form of defect.

6.1 Nucleation and Growth

For the purposes of this discussion, we consider only homogeneous nucleation as a means of promoting crystal growth. Non-homogeneous nucleation is based on the presence of impurities or foreign materials in the melt. We have assumed that the melt is pure and does not contain significant numbers of impurities that could lead to crystal nucleation sites. Nucleation sites are those locations that are preferential to the formation of crystals; they are small regions with ordered arrangements of

J. J. Nelson, *Precision Lens Molding of Glass: A Process Perspective*,
Progress in Optical Science and Photonics 8,
https://doi.org/10.1007/978-981-15-4238-1_6

atoms that provide the basis for crystal growth. Crystal growth would not occur if nucleation sites did not exist.

Upon cooling from its liquid state, the composition will naturally form a crystal structure when thermodynamics drive the system. Nucleation sites are formed in specific and limited temperature regimes where sufficient energy is available to overcome both dynamic and thermodynamic resistance to the formation of ordered structures. Devitrification of a melt requires that a significant number of nuclei are formed, and that the creation of crystals from those nucleation sites proceeds at a measurable rate.

According to Tammann, Volmer, and Weber, the rate at which nucleation sites are formed as a thermodynamic function can be stated as [1]

$$I = Ae^{(-\Delta H/kT)} \tag{6.1}$$

I = nucleation frequency
A = constant
ΔH = max change in free enthalpy
k = Boltzmann's constant
T = absolute temperature.

This shows that the formation of nucleation sites is greatly dependent on temperature and change in energy (enthalpy) of the system. This condition is met with most all melts that transition from liquid to solid phase. The time that the melt spends near the preferred temperature regime will determine the growth rate of crystals on the nucleation sites.

As the number of atoms oriented at a specific location increases, they gain structure and stability to eventually form a nucleus. Varshneya [2] has shown that the work required for nucleation to occur can be given by

$$W = \frac{16\pi\sigma^3}{3(\Delta G_x)^2} \tag{6.2}$$

W = work of nucleation
σ = surface energy of the liquid/crystal interface
ΔG_x = change in free energy for transforming a unit volume of liquid to a unit volume crystal (i.e., free energy of crystallization).

This equation again shows the need for sufficient energy levels to overcome internal barriers to formation, and reveals that the work required for nucleation to occur is simply a ratio of surface tension at the liquid–crystal interface to the energy available to perform the transformation.

The work of nucleation, W, can also be thought of as the thermodynamic barrier to creating ordered arrangements of atoms. Therefore, when the surface energy is very low, the thermodynamic barrier is very small, and only the dynamic (kinetic) barrier is of importance.

Viscosity is a dynamic material property often used to describe the behavior of liquids and vitreous materials, and it can be shown that the rate of nucleation is inversely related to viscosity of the melt (see Sect. 7.3). Considering both dynamics and thermodynamics, if we assume that $\Delta H \approx E_\eta$, and examine Eq. 7.3, then.

$$I \, \alpha \, 1/\eta \tag{6.3}$$

I = nucleation frequency
η = viscosity (Pa s)
E_η = activation energy.

6.2 Crystallization

While we use the term crystallization to describe the formation of artifacts within and on the surface of the glass, a better expression may be devitrification since the atomic arrangement does not necessarily resemble perfectly ordered locations bounded by clearly described planes, but instead are in the form of small, loosely ordered groups (crystallites) and crystalline treelike structures (dendrites).

As thermal energy is added to a crystalline solid and the substance reaches its melting temperature, the lattice elements are free to leave their near neighbor sites as translational degrees of freedom become available. The crystal structure collapses, loses its ordered shape, and transforms to a liquid state. As additional energy is added, the temperature of the substance does not immediately increase as this energy is used to remove additional elements from their lattice locations; this additional energy is known as the heat of fusion. With continued energy input, the entire substance will eventually be converted to the liquid state, and temperature of the melt will rise with increased thermal energy.

Upon cooling from its liquid state, the composition will naturally form a crystal structure when.

- thermodynamics are allowed to drive the system
- nucleation sites are available for the crystal structure to originate
- sufficient time is allowed for atoms to deposit on the nucleation sites and for crystals to redevelop.

It is important that sufficient time be provided for elements to rearrange from the amorphous liquid melt and transition to the lowest energy and preferred solid state. Time is a critical element in the formation and subsequent growth of crystalline regions, and even more so is the time spent in critical temperature zones. Just as energy input does not create an instantaneous temperature rise during melting (heat of fusion), time is required for elements to form ordered structures even as energy is removed from the system. The heat of fusion now becomes the heat of crystallization that is a function of both time and energy. As with the rate of nucleation, it can be shown that the rate of crystal growth is also inversely proportional to viscosity of the melt as.

$$u \, \alpha \, \Delta T/\eta \tag{6.4}$$

u = crystal growth rate
$\Delta T = (T_m - T)$ = degree of super cooling
η = viscosity (Pa s).

6.3 Glass Formation

As previously stated, crystals are formed when ample time is allowed for the melt to transition from liquid to solid state and when thermodynamics drive the system. Glasses are formed when both kinetics and thermodynamics define the transition and the liquid is able to pass over the melting point and continue on the same slope of declining volume, rather than the abrupt change in volume resulting from large scale crystal formation. The quickly increasing viscosity encountered during rapid cooling now becomes the driving force that prevents molecules from rearranging into a crystal structure. If the external rate of cooling exceeds the rate at which the molecules are ordered into the crystal structure, the melt falls out of equilibrium and forms a disordered, amorphous solid.

From Sect. 6.2, three conditions are needed for the crystal structure to form. To avoid devitrification, the molding process should be designed so that any or all of those conditions are not satisfied. In other words, the molding process should:

- Cool the melt sufficiently fast to bypass thermodynamic effects
- Avoid extended periods of time where temperatures are preferential to the creation of nucleation sites
- Move quickly past temperature regions that encourage crystal growth.

In order to form a clearly amorphous substance, the cooling rate must be rapid enough to force the melt into a higher energy state than it would prefer [1]. When the melt undergoes rapid cooling, movement of the molecules is hindered by the rapidly increasing viscosity and they are unable to sample nucleation sites within the time allowed by the cooling rate. They are incapable of situating in the desirable low energy locations and become frozen in a thermodynamically unstable state.

The external cooling rate affects both the optical and physical properties of the glass, while the internal rate of change is defined by the composition of the material and forms the viscosity characteristics of the melt. Slow cooling will be favorable for crystallization to occur, so rapid cool rates are used to ensure homogeneous formation of the glass. However, rapid cooling rates also create high stress levels in the solid, which contributes to stress birefringence in the final optic.

In many glass molding processes, the glass is not taken to its melting point and devitrification is not of great concern. But situations exist where higher temperatures are needed and recrystallization may be a concern. Some examples are when:

- Reflow of the optical surface is used to remove or reduce surface defects
- Sharp corners are molded that require lower viscosity to fill voids
- Molding certain "short glasses" that require higher temperatures and lower force
- Temperature control is not accurate, and the system offset brings regions of the preform near the melting point.

6.4 Inherent Stresses in Preforms

All preforms, whether ground and polished or formed from molten glass are subject to stresses that may affect their behavior during the molding process. Preforms must be handled through incoming inspection, metrology, cleaning, and storage, all before being used in molding. During the molding process they are subject to uneven heating that gives rise to thermally induced stress, as well as various types of energy inputs that may interact with intrinsic stresses of the material.

Ground and polished preforms are generally prone to surface stress from fractures and sub-surface damage that are an expected result of loose abrasive finishing. These damage sites serve as stress concentrators which if acted upon, can cause catastrophic damage to the preform. Once the preform has chipped or fractured, nothing can be done to salvage the material and it must be discarded. If the damage occurs during the molding cycle, other components in the tooling package could be affected, resulting in the removal and repair of the entire tool set.

Glass gobs and dropped spheres (preforms formed from melted glass) suffer mainly from internal stress formed during rapid cooling of the glass after creation of the final geometry. Sub-surface damage is not found on the preform since the surface has been reflowed and formed by surface tension. A compressive stress is formed on the preform (see Sect. 7.1), but this is generally not a factor during the molding process [3].

Rapid heating from RF induction systems is more of an issue than with IR or conductive systems since heat generation is so much quicker with an RF system (see Sect. 2.1.3.2). While heat transfer to the preform is still regulated by conductive properties on the tooling package, the tooling can be designed with materials that draw magnetic flux to the molding surface, which results in large thermal gradients at the point of contact between the mold die surface and preform. Preforms that are free from sub-surface damage work best with RF induction systems.

6.5 Stresses Formed During Molding

Stresses can be input to the glass article during molding through several possible scenarios:

- Rapid transition from liquid to supercooled liquid state (see Sect. 5.4.1). While it is desirable from a financial perspective to design process cycles as short as possible, quality of the finished optic may be compromised when processes cool the molded glass too quickly. Faster cooling results in solidification of the glass at a higher energy state, which translates to higher internal stress. Also, processes that maintain pressure on the molded glass well below T_g contribute to the magnification of stress in the finished article.
- Stress from heating and cooling. Whenever vitreous material is taken above the transformation temperature and atoms are given mobility to rearrange and relocate, it is almost guaranteed that some physical and optical properties will be altered due to the change in thermal history (see Sect. 5.4.1). Molding processes do not provide enough time for atoms to arrange to an equivalent low stress condition, and since polished preforms are made from fine annealed glass that is already in its lowest possible energy state, stress will be added to the finished optic.
- Stress induced by forces encountered from abrupt changes in the flow of material (diffractives, sidewalls, etc.). A solution to relieving these stresses may be through holding a constant strain on the optic and allowing the atoms in the glass to rearrange at a lower energy state [4].
- Stress caused by rapid pressing out of the glass, also known as shear thinning (see Sect. 7.9). Shear thinning is evidenced as a localized lowering of viscosity, which can correlate to an increase in temperature due to the rapid movement of molecules within the material. As temperatures decrease and these areas begin to cool, they are subject to the same increase in stress as the bulk material, although the effect is now localized on the glass surface.
- Stress at the surface as a natural function of glass molding (see Sect. 7.1). While this stress does not generally produce deleterious effects, it is nonetheless an added stress on the surface of the glass article.

While some may advocate for the use of annealing processes as a means to relieve stress after molding, annealing poses significant risk of altering the precision optical surface as molecules are allowed to rearrange to a lower energy state. In some cases, this reordering of atoms is enough to cause an in-spec surface to fall outside of customer requirements, with no remedy but to discard the optic.

References

1. W. Vogel, *Chemistry of Glass* (The American Ceramic Society), pp. 212–214 (1985)
2. Varshneya, *Fundamentals of Inorganic Glasses*, pp. 43–54 (1994)

3. J.J. Nelson et al., *Chalcogenide material strengthening through the lens molding process.* SPIE Vol. 9822, 4/2016. ISBN 9781510600638
4. J.J. Nelson et al., *Precision lens molding of asphero diffractive surfaces in chalcogenide materials*. SPIE Vol. 9633, 96331L, 10/2015

Chapter 7
Molding Processes

Just as there are many paths to success, there are no single set of parameters to define a molding process. The best process is the one that considers the many options available and adopts best practices to successful completion.

7.1 Background

The molding process has a unique effect on the properties of both oxide and chalcogenide materials, and is the one shaping process where material is not removed from the optical surface. Most other means of creating optical quality surfaces do so by starting with a block of material having a larger volume than the desired finished component. Material is then removed through repeated steps until the final surface quality is achieved. A fundamental consequence of this type of processing is that surface imperfections are added to the finished object by the subtractive nature of the operation, a consequence of grind and polish processing known as sub surface damage, which serve to weaken the overall structure (Fig. 7.1).

Single point diamond turning is another technology commonly used to produce high quality optical surfaces, and especially those having unique features such as diffractives or freeforms. As with grinding and polishing, each step rips material from the substrate, leaving behind a pattern of damage, some visible and some latent. This latent damage is commonly referred to as sub surface damage (SSD). The end result is a surface that is of optical quality, but also contains micro fractures and damage sites that weaken the overall structure.

Molding begins with a preform having either similar or exact volume of material as the finished article, and creates optical quality surfaces by heating and stretching the glass to the desired specifications. Sub surface damage is not inflicted on the material, which results in a stronger surface. In addition, the molding process creates

J. J. Nelson, *Precision Lens Molding of Glass: A Process Perspective*,
Progress in Optical Science and Photonics 8,
https://doi.org/10.1007/978-981-15-4238-1_7

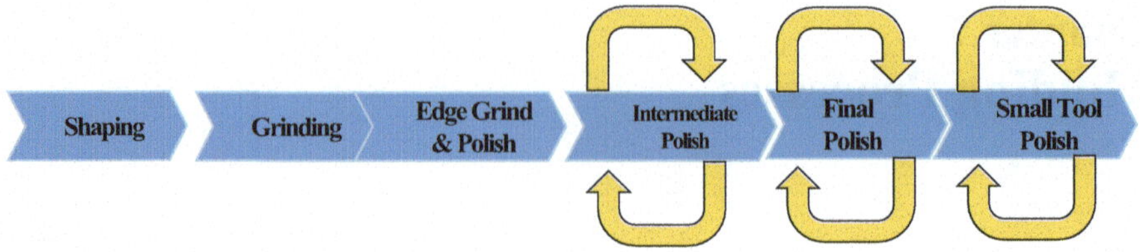

Fig. 7.1 General diagram of subtractive processes

a compressive stress on the exterior of the glass, which also serves to increase fracture toughness of the component.

Fracture toughness measurements can be influenced by both surface quality and residual stress in the material. Studies relating subsurface damage to fracture toughness have been published for oxide glasses [1], and a recent experiment performed with chalcogenide glass demonstrates the effectiveness of molding processes for improving material resistance to fracture [2]. It is commonly understood that the presence of surface fractures has a profound negative effect on certain physical properties of a material, and virtually all manufacturing processes that are subtractive in nature will produce fractures. In this experiment, a series of samples produced by various molding processes were compared against ground and polished samples of the same composition and batch.

The molded samples demonstrated no measurable variation in strength due to position, which indicates consistency over the entire surface. Also, the molded samples revealed as much as a 35% increase in fracture toughness over those parts that were ground and polished. Even though the glass used to make the ground and polished samples had been slow annealed by the manufacturer and therefore held the lowest possible internal stress, each of the molded samples outperformed the ground and polished windows (Table 7.1).

Table 7.1 Fracture toughness results for various manufacturing processes

Group	Mfg. process	Load (N)	d_{avg} (m)	K1C (MPa $m^{1/2}$)
1	Ground and polished	9.806	1.188E–04	0.2117
2	Std molding process	9.806	1.194E–04	0.2730
3	Reduced force molding	9.806	1.196E–04	0.2866
4	Impulse control, relaxation	9.806	1.193E–04	0.2848

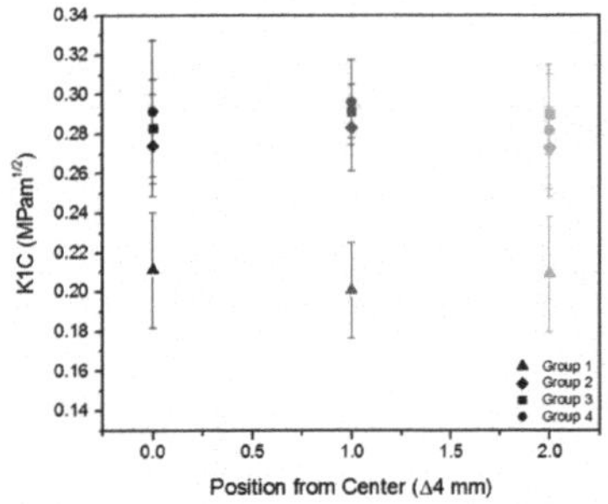

The presence of surface compressive stress in the molded samples is a natural result of the molding process and contributes to their higher measured values for fracture toughness. Upon cooling glass from the relatively high temperatures encountered during molding, the outer surface of the glass is first to reach the transformation temperature, which creates a compressive surface stress as the interior supercooled liquid continues to shrink while the outer surface is effectively frozen. This is essentially the basis for tempering of glass to improve its strength. In addition to the presence of surface fractures, the ground and polished samples also lack the compressive surface stress produced by molding processes.

Whether molding oxide or chalcogenide materials, the molding process produces benefits that are often unstated and unrecognized at the onset. But as repeated testing has shown, molding is one process that contributes in a positive way to strengthening its finished components.

7.2 Preform Selection

With few exceptions, low and mid spatial frequency errors on the preform will be transformed and reshaped during the molding process, and therefore, specifications for preforms are not especially stringent. Spatial frequency errors consistent with surface figure, irregularity, mid spatial frequencies, and slope errors will all reflect the quality of the mold tools after pressing. Surface roughness of ground and polished preforms will generally improve after molding; however, it will not replicate the surface roughness attributes of the tool except in cases where the lens is molded at very high temperature and pressure. This is because the glass does not normally achieve liquid flow during molding, and therefore the material does not reflow into vacancies on the glass surface, nor does it attain the fire polished appearance of melted glass or hot melt gob preforms. Studies have been published demonstrating improved surface roughness for molded articles [3], but the extent of this effect is dependent on the temperature and forces applied during the molding process.

Surface quality of the preform is the key characteristic that will not improve during molding, and may in fact worsen during the molding process. This is because the surface area of the preform is always being increased, or stretched, during molding.

Table 7.2 General guide to preform selection

Lens Geometry	Preform Type			
	Ball Preform	Plano Plano	Lenslet (Plano - Convex)	Lenslet (Bi-Convex)
Bi-Convex	•			•
Equal Meniscus	•		•	
Positive Meniscus	•		•	
Negative Meniscus	•		•	
Bi-Concave		•		

And since the glass is not generally held at temperature where it can readily flow and self-heal, surface defects such as scratches and digs will grow in size. When preforms are designed to minimize the flow of glass, general degradation of surface quality may not be apparent. But in cases where the preform surface area is greatly expanded during molding, surface quality imperfections may become a problem.

The following guidelines should be considered when specifying preforms:

- A surface quality grade better than the final molded optic should be specified to ensure that the final product will meet customer requirements
- If preform geometry is that of a sphere or gob, the preform diameter (or CT) should be greater than the CT of the final lens
- If the preform is planar, the CT should be greater than the CT of the final lens
- If preform geometry is bi-convex in shape, the preform sidewall should be greater than the edge thickness of the finished lens
- If the preform has a meniscus geometry, the preform CT should be greater than the CT of the final lens.

In all cases, the curvature of the preform should be such that glass can be smoothly pressed across the surface of the tool without creating voids in the finished article. Several assumptions are present in regard to material flow that are not a problem for general applications and possess few risks. We assume that material will predictably flow from center to edge, and it does in a general sense, but this may not be true for some curvatures. Closer examination of potential problems and the knowledge to reduce pressing speeds when necessary will be helpful during the process development phase. Therefore, when molding convex or planar surfaces, the preform curvature should be stronger than the final lens. When molding concave surfaces, preform curvature should be weaker than the final lens curvature (Table 7.2).

7.3 Temperature Sensing

As noted in Chap. 2, thermocouples and optical pyrometers are perhaps the most common forms of thermal monitoring used with glass molding equipment. The important point to realize is that whenever the measurement location is some distance from the object under investigation, a temperature difference will exist between the measured value and actual temperature of the glass. Depending on the sensing device employed and its proximity to the glass surface, an offset of 50 °C may be encountered and the thermocouple temperature will overstate the actual temperature at the surface of the glass. This offset may be due to a variety of reasons, but its magnitude will be a function of temperature itself, generally increasing as the system temperature increases. The danger is obvious – the preform may actually be at a lower temperature and higher viscosity than expected and will not press out smoothly, or may even fracture under loading.

However, the important parameter for molding is viscosity, which is a function of temperature, so we need to understand the effect of this temperature offset on the viscosity of the glass. Several options exist for understanding how this difference may affect the molding process. ASTM C1350M-96 (2019) describes a standardized test method for measurement of glass viscosity that can be used to determine the actual preform temperature. Other techniques may be used with varying degrees of difficulty. One method that is both useful and simple to execute is the ball penetration method described by Varshneya [4] and others [5, 6] in open literature. This method can be applied for viscosity ranges between $10^7 \leq \eta \geq 10^{11}$ Pa s.

When using ball penetration viscometry, a preform of known material is placed on the surface of a molding tool. A hard, chemically inert spherical ball of radius "R" is placed on the preform, and the tool package with preform is heated to the desired temperature. Preforms having a slight concave radius work well since the ball will naturally rest in the vertex of the preform. Record the temperature, time, and expected viscosity. Bring the tooling package back to room temperature and measure the penetration depth of the ball into the glass. The equation given by Varshneya to calculate viscosity, and therefore temperature experienced at the glass/ball interface is

$$\eta = \frac{3\mathrm{P}}{16 * \frac{\mathrm{dh}}{\mathrm{dt}} * (h(2R-h))^{1/2}} \tag{7.1}$$

P sum of applied loads
h penetration depth
R radius of sphere.

Another means of identifying the transformation temperature (T @ $\eta = 10^{12.1}$ Pa s) is found in ISO standard 7884-8, where the change in thermal expansion is identified as the transformation point of the glass (see Chap. 5, Fig. 5.3). Of course, it is always best when the complete viscosity curve can be obtained from the manufacturer, or if the VFT Coefficients (Vogel, Fulcher, Tammann) are provided. The VFT equation has been shown to provide a good mathematical description of the viscosity behavior for most common glasses over a reasonable temperature range, and is defined as

$$\log\eta(T) = A + B/(T - T_0) \tag{7.2}$$

η(T) viscosity
T_0 starting temperature
A and B are VFT Coefficients of the material.

Glass manufacturers normally provide values for key points on the temperature-viscosity curve to give an indication of how the material will respond to changes in temperature and the approximate temperature that is needed for molding. Many molding processes occur near the Littleton Softening Point. Knowing just one point on the curve or knowing points far from the molding temperature is insufficient for process development; at least one other data point is needed near the desired temperature to ensure a good starting point for process development.

Glass viscosity can also be modeled after the Arrhenius-like expression

$$\eta = \eta_0 \exp E_\eta / kT \leftrightarrow \log\eta = \log\eta_0 + E_\eta / kT \tag{7.3}$$

η viscosity
E_η activation energy
k Boltzmann's constant
T temperature.

Some useful and commonly used reference points for glass are (Fig. 7.2):

- Working Point → T @ $\eta = 10^3$ Pa s
- Littleton Softening Point → T @ $\eta = 10^{6.6}$ Pa s
- Annealing Point (T_g) → T @ $\eta = 10^{12.1}$ Pa s
- Strain Point → T @ $\eta = 10^{13.5}$ Pa s

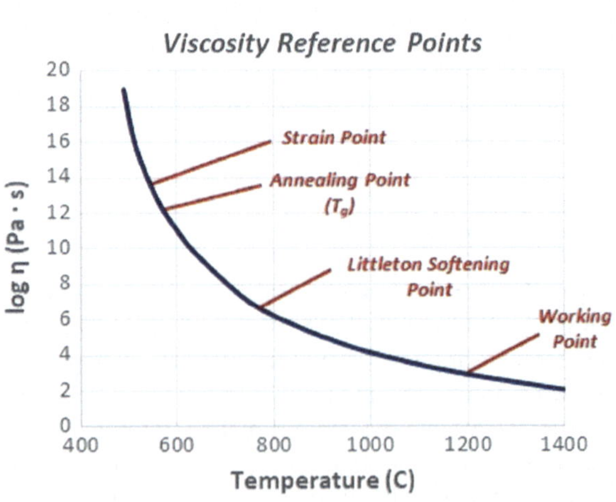

Fig. 7.2 Viscosity reference points of interest

7.4 Vacuum Assisted Molding

Once the desired molding temperature is determined and the appropriate machine settings are understood, molding process development begins. Creating a vacuum environment is a common means for removing oxygen from the atmosphere, however, machine cycle times will always be minimized when vacuum conditions are used sparingly. This is because convection processes that are useful for providing uniform heating are absent in a vacuum environment since few molecules are present in the chamber atmosphere to transport thermal energy. As a result, processes that benefit from maintaining a vacuum environment throughout the entire heating cycle are limited.

When needed, vacuum can be useful in the molding process after a nitrogen purge of the chamber, and once the process temperature has been achieved and the mold system is in soak mode. Any remaining oxygen or inert gas can be removed from the chamber to reduce the possibility of entrained gas pockets forming on the molded glass surface.

Constant vacuum through the heat and soak cycles may be used in cases where volume considerations require a ball preform whose curvature is just slightly smaller than that of the lower concave tool, with concern that as the glass presses out along the surface, the glass may seal off the lower surface and trap entrained gas, thereby producing a void on the surface. Glass flow is considered to move from the center towards the outside diameter, which is true on a large scale, but may not always be true over small regions. This is especially true when molding aspheres that are not well behaved and contain surface slope reversals (Figs. 7.3 and 7.4).

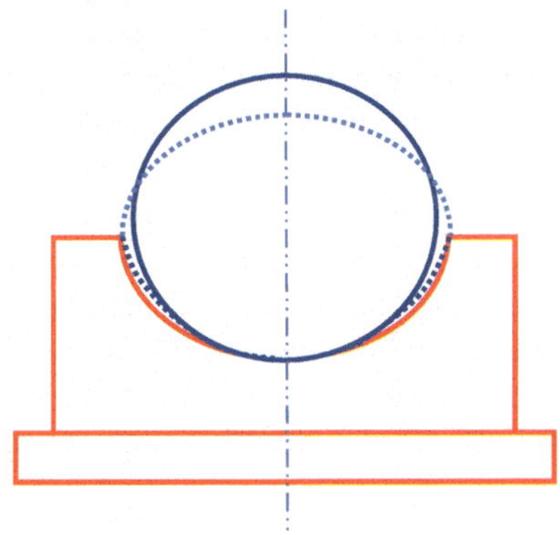

Fig. 7.3 Preform touching edge of mold tool

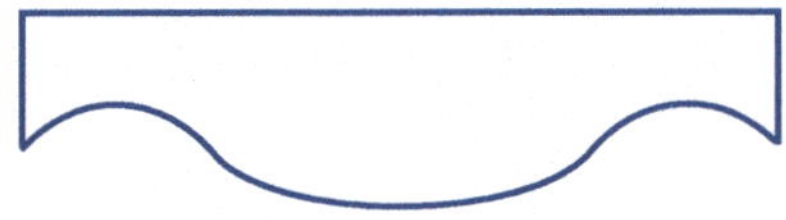

Fig. 7.4 Exemplar asphere with inflection surface

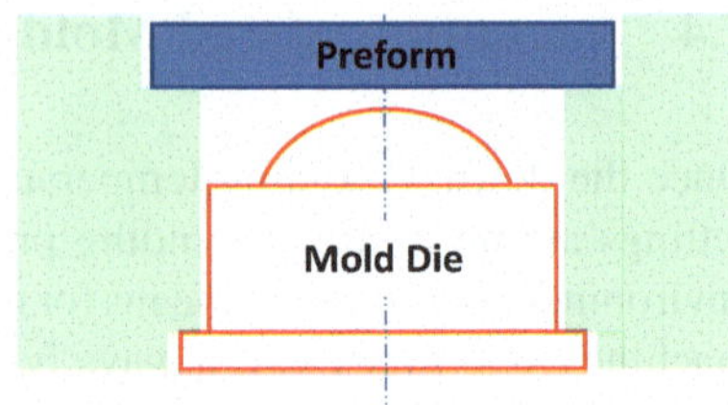

Fig. 7.5 Molding negative lenses with planar preform

Vacuum molding may also be useful when molding negative lenses from a planar preform. The planar preform is caused to reside on a ledge above the lower convex tool, and vacuum is used in the hopes of removing entrained gas between the preform and lower tool. In reality, only a minimal amount of gas is removed, but oftentimes this is sufficient to produce a void free optical surface on the lower tool. Any gas that is trapped in the lower cavity is pressed to the circumference as long as the process allows glass to first contact the vertex of the lower tool and slowly press out to the edge. If pressing speeds are too great, chances are high that an incomplete surface will be molded due to glass rolling over the lower tool instead of flowing across the tool surface (Fig. 7.5).

7.5 Finished Lens Molding

Finished lens molding (FLM) and free diameter molding (*PLM,* see Sect. 7.5) can both use the same types of processes. Force control molding, position control molding, constant flow molding, and stress relaxation techniques are all within the possibilities of FLM and PLM processing. Finished lens molding, also known as mold to diameter, is the condition where no additional processing is required after molding; the lens is molded complete with the correct optical features and specifications, diameter, center thickness, and special features such as mounting flanges or datums.

Examine Fig. 7.6 and notice how the convex surface transitions to a short horizontal surface, and then a sharp transition to the edge where it forms the outside diameter. The diameter, center thickness, and all optical surfaces are within tolerance. Molding with FLM processes requires careful attention to all specifications and tolerances that affect the volume of the finished lens since the preform will of necessity be held to the same or tighter volume controls. Release of the diameter

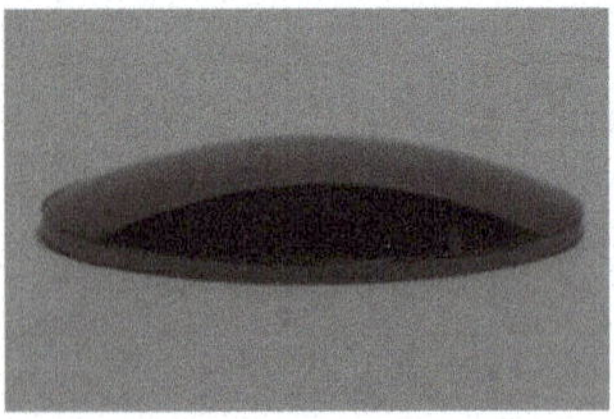

Fig. 7.6 FLM of chalcogenide meniscus lens

Table 7.3 Exemplar volume tolerances of meniscus lens and preform

CA surface 1	25.150	CA surface 2	23.800
Lens diameter	25.400	Lens CT	2.750
CX radius 1	30.566	CC radius 2	90.811
Chamfer S1	0.20	Chamfer S2	0.20
V_{total} $(mm^3) =$	2796.07	Ball $Ø_{total}$ (mm) =	17.479
V_{max} $(mm^3$ $=$	2800.65	$Ø_{max}$ (mm) =	17.489
V_{min} $(mm^3$ $=$	2781.69	$Ø_{min}$ (mm) =	17.449

from the mold also needs to be considered and will be affected by the force and temperature profile selected. Typically, lower forces and/or lower temperatures are used in FLM processes to avoid pushing too much glass into the sidewalls of the mold die.

Consider a meniscus lens having specifications of 25.4 mm $\pm$ 0.025 diameter, 2.75 mm + 0/−0.020 CT, and 0.2 mm safety bevels (or rounded edges). Power and irregularity tolerances are assumed to have minimal effect on the final volume. The total allowable variation in volume is 14.38 mm^3, and therefore, a spherical preform may be specified with a total allowable diameter tolerance of 40 microns when specifying finished lens molding processes. Stipulating a preform with this tolerance or slightly less ensures that all differences in volume of the finished lens will satisfy customer requirements (Table 7.3).

FLM processes are not always defined by the sharp edges on the diameter; processing may be used to create rounded edges on the diameter in those instances where the diameter tolerance is not tightly controlled. Under those conditions and given the previous example, the diameter tolerance could be as large as the supplier needs, so long as the minimum and maximum volume of the preform falls within the volume tolerance of the finished lens.

7.6 Precision Lens Molding

Precision lens molding (PLM), also known as free diameter molding, is the more general term for glass molded optics. PLM is the process whereby the optical surfaces are precision molded, but post processing is required to meet customer specifications. Post processing is considered to be any additional operation to the physical dimensions of the glass such as centering, creation of sag dimensions, segmenting, and the like. Coating is not considered a post process since every lens should have a coating of some type applied after molding to reduce reflection losses. Note the smooth, rounded edges of Fig. 7.7 where the glass was allowed to free flow and form the outside diameter.

Given the conditions of a 25.4 mm diameter lens from the previous example, the diameter tolerance on a spherical preform could be as large as the supplier needs, so

Fig. 7.7 PLM of chalcogenide meniscus lens

long as the minimum volume of the preform is greater than the maximum volume of the finished lens.

7.7 Sleeve Molding

Certain applications will benefit from molding within a sleeve, either in a single cavity or multi cavity configuration. For example, assume a customer request is received for a small quantity of 12.95 mm diameter optics, but no such tooling exists in inventory. It may be possible to use a larger diameter mold die with an appropriate designed insert, which will save both cost and lead time, resulting in a better product for the customer (Fig. 7.8).

Certainly, material choice of the sleeve is important, but it also provides the designer with an opportunity to reduce or remove centration and wedge errors from the finished optic. By choosing materials that take advantage of differential expansion characteristics so that the sleeve expands slightly to fill the clearance needed for room temperature assembly, a higher precision article may be molded. The same principle can be used on single cavity tooling where existing tooling is designed for a larger diameter than a current requirement. Whether in single cavity or multi cavity configuration, sleeve molding can help reduce both costs and lead time and is a cost effective means of making use of existing tool inventory.

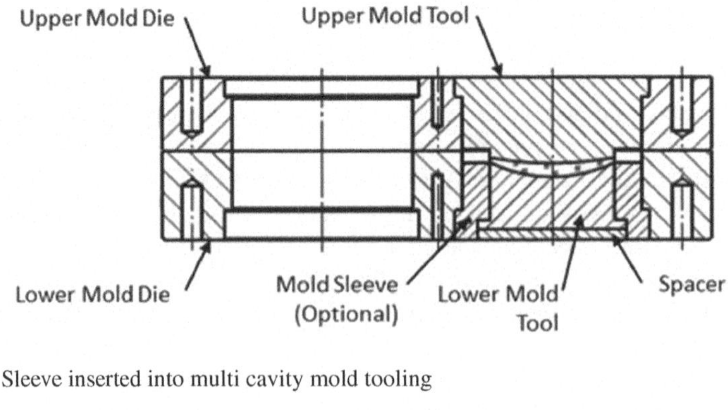

Fig. 7.8 Sleeve inserted into multi cavity mold tooling

7.8 Position Control Molding

When position control molding is selected, the pressing ram moves to a preset location in space, regardless of the force required to achieve its position. Position control should not be employed until a sufficient number of dry cycles are run and the machine has been able to achieve some level of thermal equilibrium. All of the components in the molding machine will increase in temperature over time, and as a result, most will expand in length. This results in a changing of the "zero point" of the machine readout relative to a location in space. Three or four dry cycles are usually sufficient for most components to thermalize.

This option is not used when molding to a closed mold position; that is, when the upper and lower mold dies come into intimate contact with each other and form a positive mechanical stop. If the entered location on position control is beyond the contact point of the mold halves, force will continue to build as the machine moves to satisfy the input position command, and significant damage to the machine could occur.

Position control molding is useful when tight control of lens center thickness is required and positive stops in the tooling are not available, or when abrupt changes in the optical surface are encountered, or when slower displacement speed is needed for certain regions that make up a small part of the overall surface area (see Sect. 7.13).

7.9 Force Control Molding

Most molding processes run in force control mode; that is, a maximum pressing force is set by the operator, and the ram contacts the glass and continues to press out the surfaces until the maximum force value is reached. As the glass is stretched across the mold tool and surface area expands, the viscous force that resists the downward force of the pressing ram increases. For some time after reaching maximum force, displacement may continue as the glass slowly flows across the tool surface until the molding cycle time set point is reached or a hard stop is encountered by the tooling.

From Fig. 7.9, notice how glass displacement continues long after maximum force is reached, and how the rate of displacement continues to decline quadratically as the surface area of glass increases. In this instance, movement of the ram continues until the force is removed.

Just as rapid penetration followed by a gradual decline in displacement rate is shown on a macro scale, the same is true on a much smaller scale. As the ram makes initial contact with the glass and encounters a very small surface area, it moves almost unimpeded by the viscous force of the glass resisting change, and as the contact area increases, displacement rate of material and ram speed decrease. This initial contact produces very high pressures across the glass surface and causes glass to be displaced at a very high rate compared to bulk flow rates. So we see two forces at work during

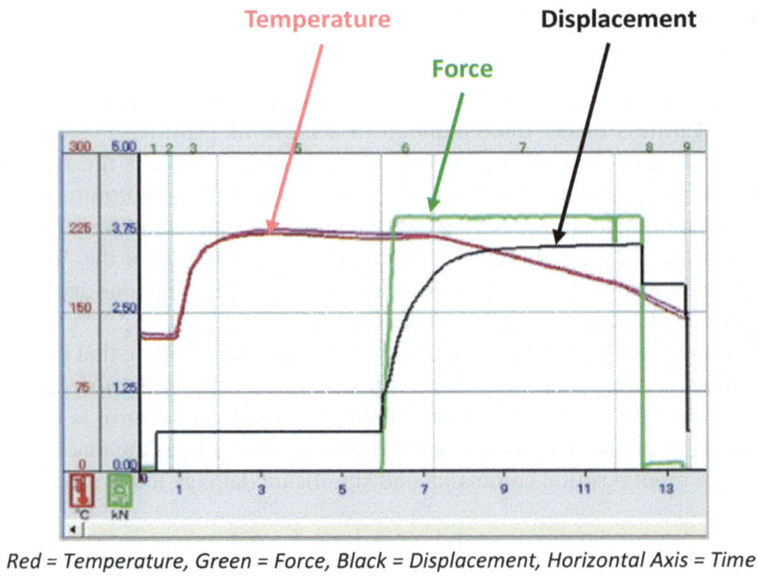

Fig. 7.9 Exemplar process for force controlled molding

press out; first is the impact force based on the velocity of the pressing ram at first contact, and second is the maximum press force as entered by the operator.

As the heated glass is caused to move at a high rate of speed, sufficiently high shearing rates may cause it to depart from Newtonian behavior [7]. This shear thinning is evidenced by a localized lowering of viscosity, leading to increased phase separation that may produce problems when working the glass. Not only does the viscosity become non-linear at high shearing rates, shear stress also increases rapidly as a result of inadequate rates of stress relaxation.

Results of high shear rates are shown below for some chalcogenide materials. In this instance, the pressing ram encountered the glass at a relatively high rate of speed, resulting in a damaged region that extended around 5 mm in diameter from center (Fig. 7.10).

As shown by Fig. 7.9, several minutes of press time were needed to fully form the full 25.4 mm diameter, but a closer look at Process A in Fig. 7.11 reveals that the central 5 mm diameter is formed in only 0.273 s with a ram speed of 0.128 mm/s and volumetric displacement rate of 9.45 mm^3/s. Even when speeds were slowed to 0.047 mm/s as in Process B, damage still occurred even though formation of the central 5 mm slowed to 0.745 s at a volumetric rate of change of 4.51 mm^3/s.

Investigation of material composition in both the affected region and undamaged substrate are shown in Figs. 7.12 and 7.13. Note that no decomposition of the material occurred, nor were any areas found with higher than normal concentration of any element. The effect was one of shear thinning that resulted in a cloudy region being formed on the molded glass surface. This cloudy area appears to be the result of

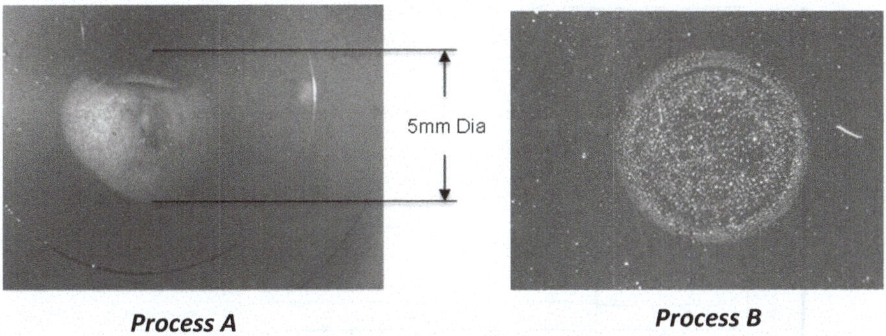

Fig. 7.10 Damaged region at center of lens from initial contact with mold tool

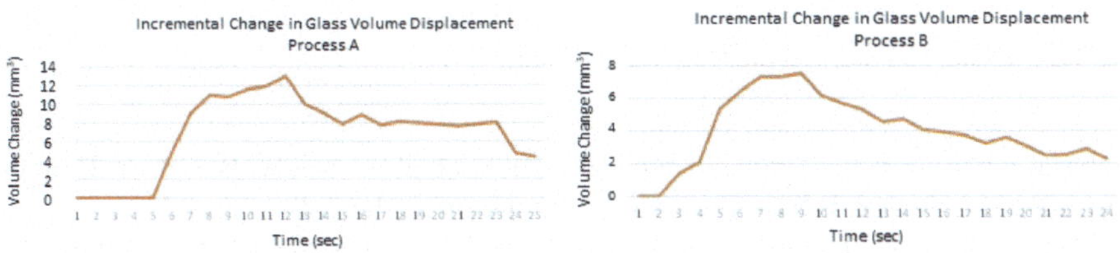

Fig. 7.11 Displacement of glass from initial contact with mold tool

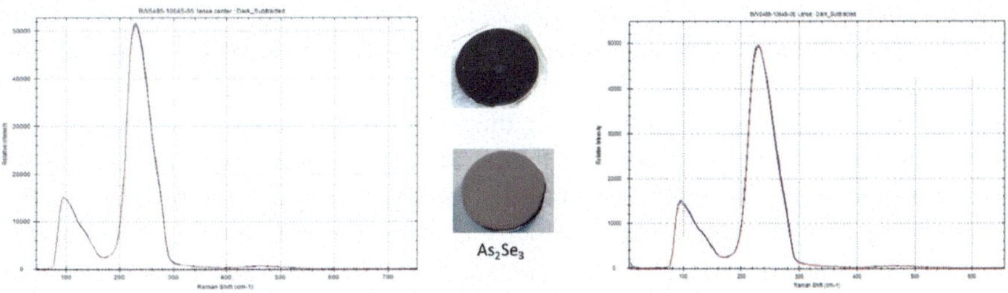

Fig. 7.12 Raman spectroscopy of damaged region (left) and virgin material (right)

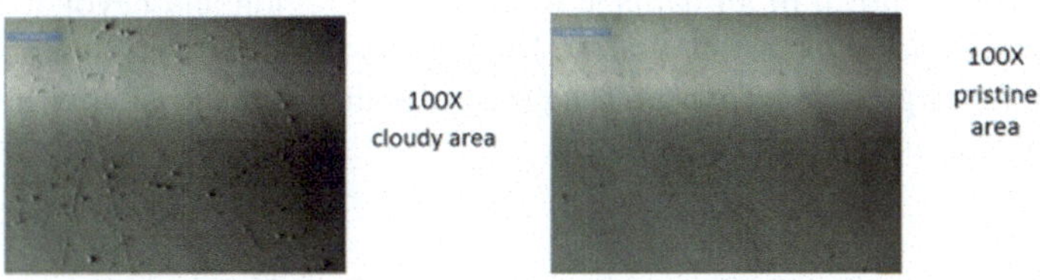

Fig. 7.13 Microscopy of damaged region (left) and virgin material (right)

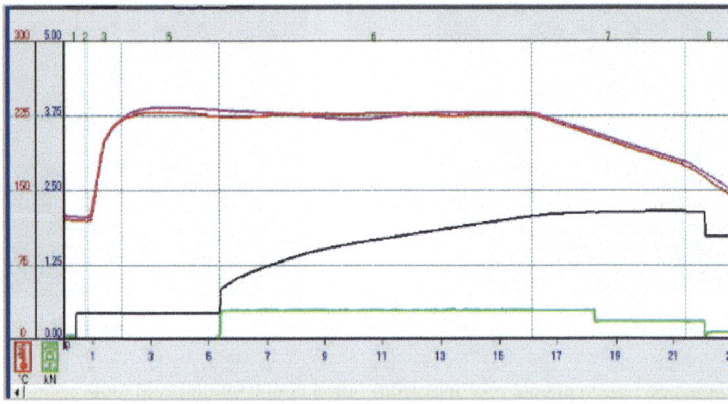

Red = Temperature, Green = Force, Black = Displacement, Horizontal Axis = Time

Fig. 7.14 Exemplar process for reduced force molding

material separation that induced the nucleation of glass beads. Based on Raman spectroscopy, the beads have the same composition as the bulk material.

Microscopy of each region confirmed this conclusion. The effects of shear thinning and phase separation resulted in micro balls of material being rolled up on the surface of the molded glass.

With many materials and process speeds, this is not an issue and damage will not be apparent. But under some conditions, and especially when working with chalcogenide glasses, care should be exercised when impact speeds are high and central molding defects are a possibility. While this effect is not generally as pronounced when working with oxide glasses, if a central cloudy region is evidenced on the molded article, then slowing the ram speed at initial contact may improve the results.

One means of lowering contact velocity is to set the maximum force to a very low number while continuing in force control mode. In one experiment, parameters were set so that the maximum press force did not exceed 2.0 kN, and as a result, the pressing cycle required 325 s to complete full displacement. The result of this change was the removal of all damage in the central region and beyond; the entire molded surface was of excellent quality. The penalty for taking this approach is that the cycle time is greatly increased, which results in lower throughput and higher overall manufacturing cost (Fig. 7.14).

7.10 Constant Flow Molding

As it has been shown that even moderately high contact velocities of the pressing ram can result in shear thinning, our thoughts must now focus on how to slow things down in the area of concern without causing a large increase in cycle time. The ideal solution is to control the flow of glass throughout the pressing cycle and produce a relatively constant flow across the tool surface rather than one that is dependent on penetration rate of the mold tool.

Constant areal displacement of glass with respect to time can be written in equation form as

$$\frac{dA}{dt} = constant \tag{7.4}$$

By rewriting this equation, we see how areal change can be controlled with machine settings

$$\frac{dA}{dt} = \frac{dA}{dx} * \frac{dx}{dt} = constant \tag{7.5}$$

$\frac{dA}{dt}$ is the areal change with respect to time
$\frac{dA}{dx}$ is the change in area with respect to penetration depth
$\frac{dx}{dt}$ is the velocity of the pressing ram.

Since the aspheric equation is also useful to describe spherical and planar surfaces and is the most general form to describe all three, we simply define the area of an aspheric surface and take the derivative of this areal solution with respect to position to arrive at the following;

$$\frac{dA}{dt} = 2\pi\sqrt{\mathrm{R}^2 - (2\mathrm{R} * \mathrm{k} * \mathrm{x}) + \left(\mathrm{k}^2 + \mathrm{k}\right) * \mathrm{x}^2} * \frac{dx}{dt}$$
$$\text{Change in area wrt position} * \text{Feed Rate} \tag{7.6}$$

Consideration of higher order terms require similar calculations. These principles were applied to the lens described in Table 7.3, and a new process was entered into the machine software (Process C) to allow glass flow at a slower and relatively constant rate, even as the surface area greatly increased. Applying these principles yielded the following results (Fig. 7.15);

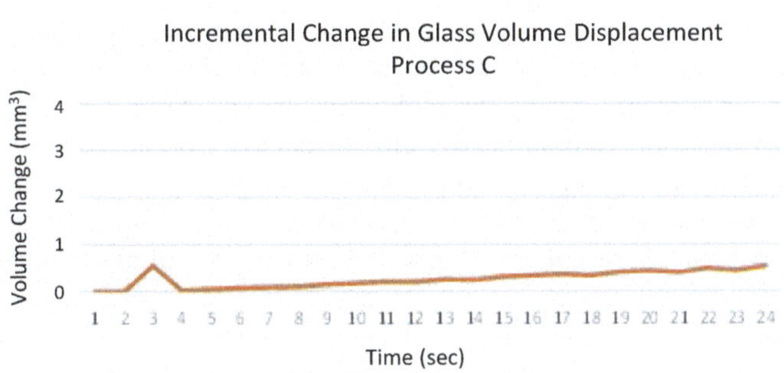

Fig. 7.15 Displacement of glass from initial contact with mold tool

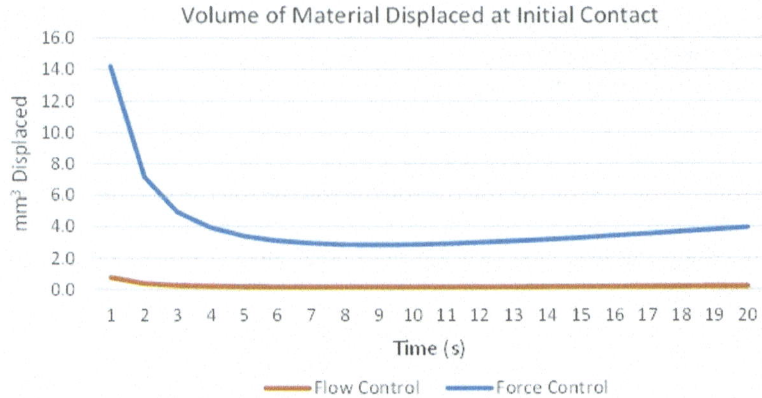

Fig. 7.16 Volume of material displaced by pressing ram

By making this simple process change, damage to the central area was completely eliminated. A comparison of the volume of glass displacement with respect to time for Process A and Process C is shown in Fig. 7.16.

By placing both sets of data on a common graph, the difference in material flow is readily apparent between the force control and constant flow processes. This type of change can be easily inserted into the molding program for any desired flow rate that is appropriate for a specific application, and the flow rate can be varied for specific radial zones. The time needed to achieve full press out will also be obtained empirically during the first couple of cycles.

An added benefit of exercising a controlled flow of material is that the glass now has time to relax and relieve internal stresses that build up during the molding process, which results in a stronger component that is less prone to chips or fractures [7].

7.11 Stress Relaxation

Stress in the molded glass is a function of several mechanisms; elastic stress, anelastic stress, and a viscosity component, each of which exists at some level for most temperatures encountered during the molding cycle [8]. Also, it is the property of viscoelasticity that induces nonlinearity into the behavior of the material and allows the rearrangement of molecules to reduce stress in the glass [9]. The fact that it seems to occur over a relatively short time period during molding compared with room temperature experiments is reasonable due to the contribution of thermal energy being input to the system.

Recall that viscosity can be thought of as internal friction, and lower viscosity is evidenced by faster response to the external force exerted by the pressing ram and less interaction between molecules. Stress relaxation is possible due to the viscoelastic nature of the glass, and relaxation time is decreased through the input of thermal energy that supports molecular movement within the material. This can be described by an Arrhenius type relationship where.

$$\sigma_{final} \approx A \cdot \sigma_{original} \cdot e^{-(t/\tau)} \tag{7.7}$$

A is a constant.

τ is the relaxation time that is a function of material properties and temperature.

Relaxation time is the amount of time required for the current stress to be reduced to 1/e of the initial stress level.

$$\tau = \eta/k \tag{7.8}$$

k is analogous to Young's modulus.

From Eq. 7.7, it is apparent that the final minimum stress level will be a function of time, viscosity, and stress history of the glass. Therefore, stress relaxation techniques that are applied at higher temperatures will experience the greatest benefits of lower overall cycle times and final stress levels. Recall that at temperatures above the transformation point of glass, additional degrees of freedom of motion become available and the glass is able to respond much more rapidly to a variety of external stimuli (Sect. 5.3). Therefore, stress relaxation should be applied at temperatures above the transformation point of the glass.

Most mathematical models predict that stress levels will approach zero as time goes to infinity, but from Fig. 7.17 and Eq. 7.7, it can be seen how higher original stress levels impact the minimum achievable stress in the finished component. As all three processes are carried out through time, they all experience a similar rate of internal stress decay and reach an asymptotic minimum whose relative value is commensurate with the initial stress level. The stress curves that possess higher initial stress values never attain the minimum values reached by the samples having lower initial stress. So when we consider reasonable and finite time intervals, we see that each sample reaches a different minimum stress level, with stress history

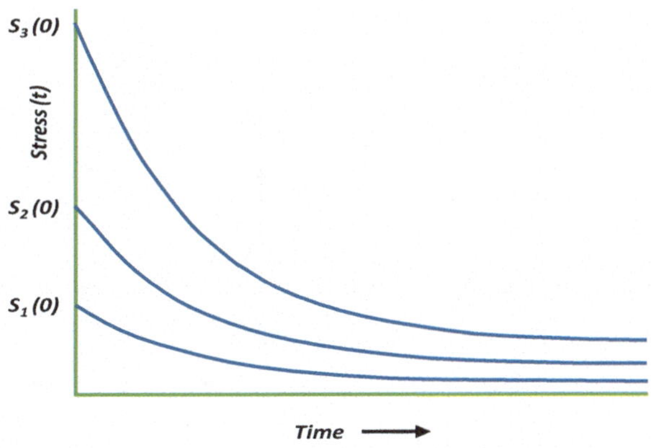

Fig. 7.17 Stress relaxation behavior of 3 samples with various stress histories

being a major factor in determining the final minimum stress. Therefore, it is equally important that stress in the glass be minimized from the beginning of the process in order to experience the lowest possible stress in the finished article.

Stress relaxation is most often accomplished through the application of a constant strain over a period of time. When a body is exposed to a constant strain (obtained by holding a constant position during molding), there is a gradual decay in stress levels as long as molecules have some freedom of movement. In effect, the process is allowing time for the glass molecules to rearrange to a lower stress condition rather than continuing to force the glass into a specific shape and structure. Once again, determination of the necessary relaxation time is purely empirical (Figs. 7.18 and 7.19).

An experiment was performed with chalcogenide glass and the lens design shown in Table 7.3 where the process was set to position control and a series of constant strain

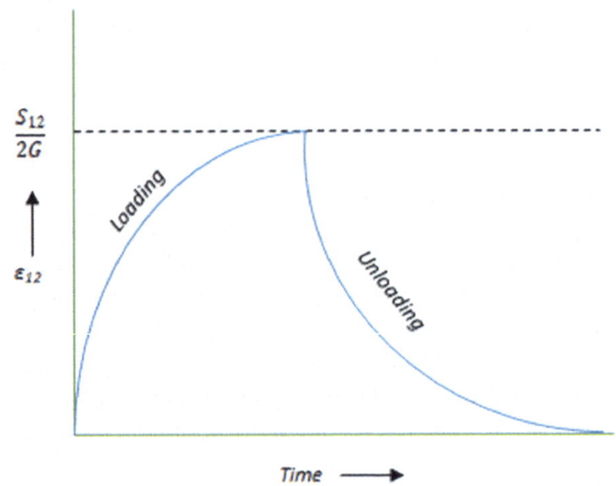

Fig. 7.18 Response after application of constant stress

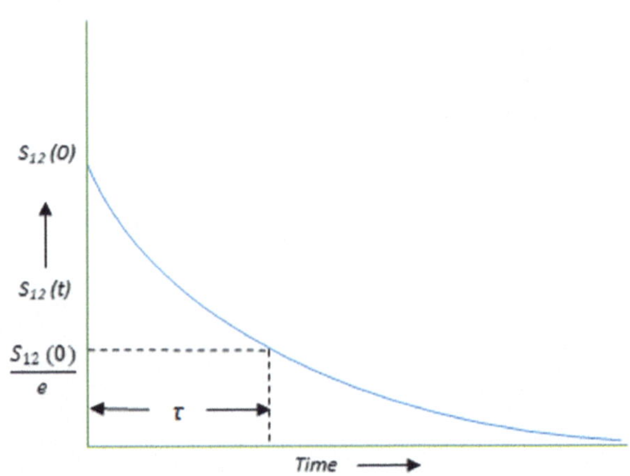

Fig. 7.19 Response after application of constant strain

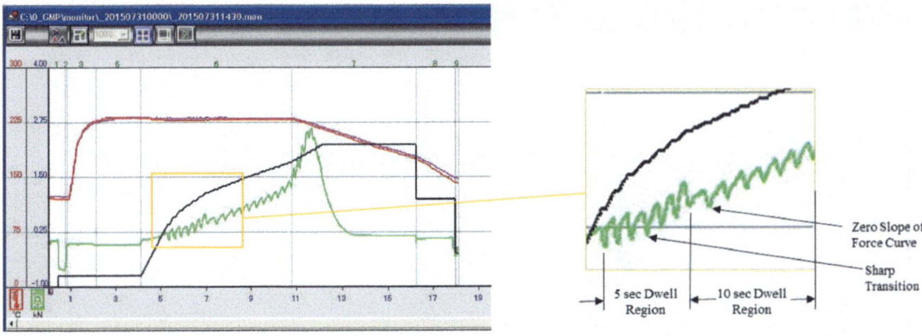

Fig. 7.20 Stress relaxation by creation of constant strain regions

regions (dwell periods) were added. The length of dwell was varied to understand the effect of time on relaxation and to understand what time and temperature settings would be appropriate (Fig. 7.20).

At early stages of the process when the pressing force and resultant stress were minimal, two second dwell and five second dwell periods were examined. Both of these dwell periods demonstrated a definite reduction of stress, but it appeared that stress was not fully minimized since the force profile made a sharp transition from decreasing to increasing once the ram resumed its motion, and the new local minimum was greater than the previous minimum. However, when a ten second dwell was examined, it appeared that the stress had relaxed to a minimum since the force profile revealed a short section of zero slope, indicating that no further reduction could be attained under those process conditions.

Subsequent regions of ten second dwell continued to reduce stress at a slightly faster rate than the five second regions, but the minimum value continued to increase. This ever increasing minimum confirms the interpretation of Eq. 7.7 and shows that higher initial stress levels require longer periods of time to relieve stress, and that once the stress is created in the glass, time and temperature are the main parameters to affect relaxation. In general, a significant reduction in stress is adequate for most applications, and the added time required to create a zero slope condition at each dwell step is not necessary (Fig. 7.21).

7.12 Insert Molding

In some cases, it is advantageous to mold the lens directly inside of a metal barrel or insert. Examples may be small diameter optics that require critical alignment to a datum on the housing; or multi element designs where one element needs to be mounted deep within the barrel; or high volume, low price point designs that could not absorb the cost of assembly.

When considering insert molding, special attention must be given to the thermal expansion properties of materials, the type of fits that are required, and stresses formed between glass and insert. The thermodynamic principles discussed in Chap. 3

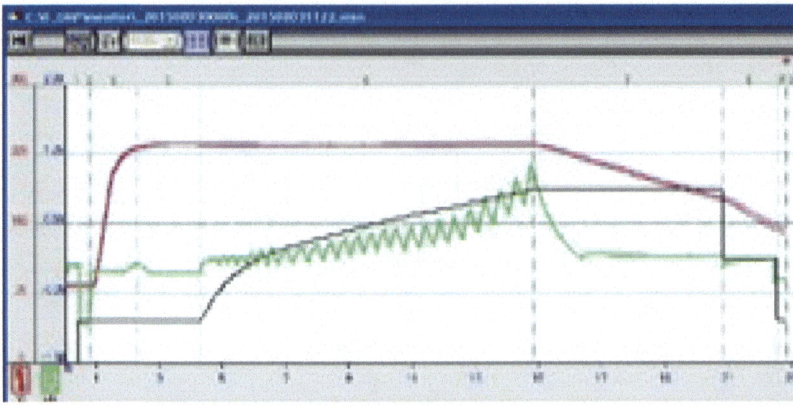

Red = Temperature, Green = Force, Black = Displacement, Horizontal Axis = Time

Fig. 7.21 Exemplar process having stress relaxation features

remain valid, but the presence of a metal barrel restricts the type of heat source that may be used.

Resistive sources should not be affected by the barrel, other than having to transfer heat through a greater thermal mass. Infrared lamps will be somewhat affected, in part by the material properties of the barrel (whether it absorbs at the emitting frequencies), and partly since the presence of a barrel obscures radiation from directly interacting with the mold tools, which will increase soak time and may change the thermal gradient between glass and thermocouple. RF sources will be most affected since the presence of any metal will affect the magnetic flux passing through the mold tools and alter the thermodynamic workings of the system.

A large amount of information is available in public domain, and careful consideration should be given to the dynamics associated with insert molding before risking damage to expensive mold tooling [10].

7.13 Molding Special Features

Just as it has been shown that rapid changes in vertical displacement cause significant changes on the radial flow of glass and have a negative effect on the glass surface, the same is also true for geometries that require step profile changes of the optical surface. When the moving glass encounters a wall that inhibits movement, as in the molding of diffractive surfaces, large amounts of force are built up as the freely flowing glass is faced with instantaneous and infinite resistance.

An experiment was performed on a chalcogenide lens described in Table 7.3, where a four zone diffractive was added to the spherical concave surface at a design

Ring	Zone Radius (mm)
1	5.60 ± 0.05
2	7.90 ± 0.05
3	9.80 ± 0.05
4	11.20 ± 0.05

Fig. 7.22 Diffractive zones and representation

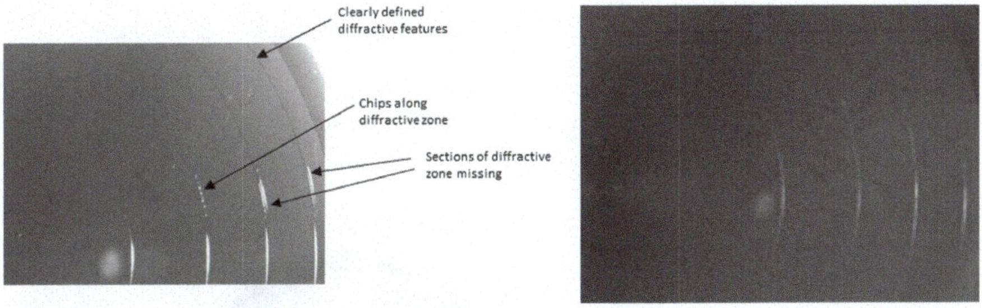

Fig. 7.23 Diffractive zones damaged (left) and fully formed (right)

wavelength of 10 microns and a step height of 5.6 microns [3]. Standard force control molding, reduced force control, constant flow molding, and stress relaxation processes were all tested on the same tooling (Fig. 7.22).

Formation of the diffractive surfaces was unsuccessful when molding with a standard force control program (4.0 kN maximum force), and while a reduced force (2.0 kN maximum force) program produced solid rings for the first zone, each of the outer zones displayed damage. Use of a constant flow program produced good results for each of the first two rings, but the outer two rings still displayed damage. It wasn't until the constant flow program coupled with stress relaxation provisions were added that solid rings without any defects were formed for all diffractive features (Fig. 7.23).

The constant flow/stress relaxation process reduced stress by over 90% when compared with standard force control processing. Diffractive features were well formed and consistent in both shape and figure when using both CTE matched tooling and non-CTE matched tools, which demonstrates the effectiveness of proper stress management (Fig. 7.24).

Another feature that is commonly molded with an optic is the addition of mounting flanges. Flanges are particularly useful for mounting lenses in a barrel, and can simplify set up when surfaces are referenced from the flange. For example, in some systems the Sagitta of a surface or location of a vertex is important to the system design. These dimensions can be permanently and accurately controlled by the use of good tool design (Fig. 7.25).

Non-rotationally symmetric surfaces such as flat regions cut into the flange is another example of special features that are helpful to the designer. It is often difficult to cause the glass to flow into sharp corners, but a high quality flat can be molded into

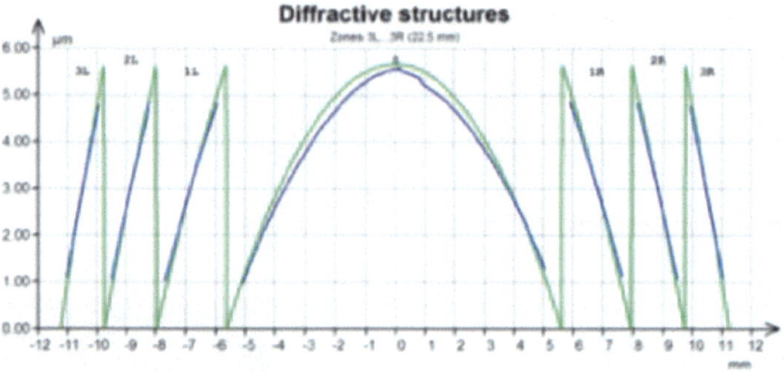

Fig. 7.24 Overlay of measured data with theoretical diffractive surface

Fig. 7.25 Chalcogenide molded lenses, with and without mounting flanges

the flange as long as a radiused transition to the flange is acceptable. The addition of a flat on the diameter of the flange provides orintation to the finished lens, which may be useful in some situations.

An extension of the constant areal flow process is to adjust the flow based on regions of concern; that is, to create radial zones where glass movement can be higher or lower based on the features of concern. This type of process may be useful for large diameter optics having only a few diffractive zones where moderate speeds can be programmed in the smooth transition regions, but then slowed over the diffractive step, and then increased again over smooth regions.

In general, consideration should be given to molding many different types of features that may not commonly be found on molded optics. With thoughtful processing and preform design, many options exist to help the designer reduce costs, improve quality, increase yields, and simplify designs and assemblies.

References

1. S. Jacobs et al., Subsurface Damage in Microgrinding Optical Glasses, vol. 73 (University of Rochester, Laboratory for Laser Energetics, LLE Review, 1998)
2. J.J. Nelson et al., Chalcogenide Material Strengthening Through the Lens Molding Process. SPIE Volume 9822, ISBN 9781510600638, 4/2016
3. J. Hu et al., Optical Loss Reduction in HIC Chalcogenide Glass Waveguides Via Thermal Reflow (Micro Optics Center, Massachusetts Institute of Technology, CLEO/QELS Conference, 2009)
4. A.K. Varshneya, Fundamentals of Inorganic Glasses (1994), pp. 219–225
5. R.W. Douglas, W.L. Armstrong et al., A penetration viscometer, in *Glass Technology*, vol. 6, no. 2 (April 1965), pp. 52–55
6. R. Brückner et al., Systematic investigation of the use of penetration viscometers, in *Glass Technology*, vol. 48, n. 1 (Jan 1975), pp. 12–18
7. A.K. Varshneya, Fundamentals of Inorganic Glasses (1994), pp. 206–208
8. Pierre Lucas et al., Sub Tg viscoelastic behavior of chalcogenide glasses... Journal-Ceramic Society Japan (July 2008)
9. H. Kadali, Experimental Characterization of Stress Relaxation in Glass (Clemson University, Jan 2009)
10. A. Symmons, B. Auz, Design Considerations and Manufacturing Limitations of Insert Precision Glass Molding, SPIE vol. 8489 (LightPath Technologies, 2012)
11. J.J. Nelson et al., Precision Lens Molding of Asphero Diffractive Surfaces in Chalcogenide Materials. SPIE Volume 9633, 96331L, 10/2015

Chapter 8
Applications

Combined with technology advancements in supporting industries, glass molding has found application in everyday devices from cell phones and computers to specialized applications in military and life sciences. Quality results only come from good processes; inferior equipment can produce excellent quality optics when supported by technical understanding and good process control.

8.1 Background

8.1.1 Technology Impacts

Advancements in equipment technology and manufacturing processes have dramatically increased over the past four decades. With accuracy and precision of both tool manufacture and thin film coatings showing exponential improvement, molded glass optics have entered consumer and industrial markets to provide high quality components at low cost. Whether for high or low volume applications, PLM technology has helped meet market demands to transform optics manufacturing.

Applications for molded glass optics have increased in recent years due to several global market phenomenon:

- Increased asphere production. Incorporation of aspheres in a design generally results in fewer lenses in the system, but conventional fabrication techniques rely on the production of individual lenses that drive up costs. Molding allows processing of multiple aspheres in a single cycle, thereby reducing operating costs.
- Miniaturization of optical elements and reduction of system size and weight. Conventional grind and polish equipment is not designed to produce elements having small diameter or radius of curvature. For infrared applications that rely on diamond turning technology, smaller elements cause difficulty securing the glass to

J. J. Nelson, *Precision Lens Molding of Glass: A Process Perspective*,
Progress in Optical Science and Photonics 8,
https://doi.org/10.1007/978-981-15-4238-1_8

the vacuum chuck that results in low yields and increased waste. Molding provides a solution to these problems with the ability to produce optics having very small diameter and curvature (less than 1 mm), as well as much larger surfaces (over 100 mm diameter) when the correct equipment is used.

- Precision pressings. Preforms of the exact or slightly greater glass volume result in lower cost of materials than for grinding and polishing processes that begin with a much larger volume of glass. This is especially relevant for applications that require high numbers of annual units.
- Expansion of moldable glass types. An expanded glass chart provides the designer with more choices for the optical properties that are needed for high performance designs, while facilitating the use of molding technology.

8.1.2 Glass Types

Vitreous silica is the basis for many commercial glasses used for molding, but with a softening temperature of nearly 1700 °C, fused silica is rarely considered for molding applications except in extremely rare cases and with specialized equipment. It is frequently used for optical fibers, astronomical mirrors, and as thermal or electrical insulators.

Borosilicate glass is used for precision glass molded lenses where low thermal expansion and high resistance to chemical attack are required. Low expansion and higher thermal diffusivity values result in lower distortion when cooling. Borosilicates are also considered to be crown glasses that possess lower refractive index and low to moderate dispersion properties which are useful for many optical designs, especially for correction of secondary color aberrations.

Aluminosilicate glass is known for higher values of Young's modulus and high resistance to chemical attack. Applications are common where higher operating temperatures are encountered, or where higher than normal stress is anticipated in the design.

Lanthanum oxide glasses typically possess higher hardness, higher transformation temperature, and useful refractive index and dispersion properties. Higher index enables shorter overall system length and lower spherical aberration. Higher dispersion properties are useful in many optical designs for both consumer and high end optical systems (Fig. 8.1).

Many variations in stoichiometry are possible, with each element and compound adding its own unique character to the final properties of the glass. The ever increasing availability of moldable glasses should continue to help in the growth of PLM technology for years to come.

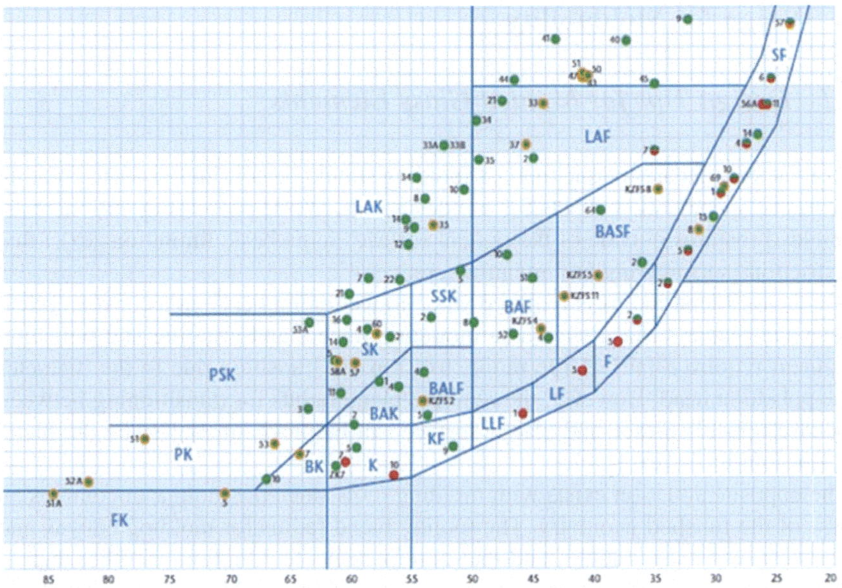

Fig. 8.1 Refractive index/dispersion map of SCHOTT glasses

8.1.3 *Optical Surfaces*

Planar, spherical, aspheric, diffractive, cylindrical, acylindrical, arrays; all of these surfaces can be molded with careful planning and the correct equipment. Greatest benefits from PLM processing are realized when producing surfaces having a distinct orientation or axis of rotation (aspheres, acylinders, freeforms, and arrays) and those with special features that can be molded (diffractives, some non-rotationally symmetric geometries). Planar and spherical surfaces can be easily molded, but lenses having only these features that lack a distinct axis of rotation are best produced by traditional methods (Fig. 8.2).

Surface Description	Manufacturing Cost	Comments
Bi-concave; aspheric		Very high risk
Bi-concave; spherical - aspheric	High	Post processing may be required
Meniscus (spherical or aspheric)		
Bi-convex; aspheric - aspheric		Slightly more expensive than spherical - aspheric
Bi-convex; spherical - aspheric	Low	
Planar - Convex (sphere or apshere)		Rotationally symmetric

Fig. 8.2 Manufacturability guide for PLM operations

8.2 Design for Moldability

8.2.1 Design Tips for Molded Glass Elements

Several factors that are unique to glass molding should be considered when creating drawings or specification sheets for molded optics. While processing options exist for molded features outside of the optical surface, some basic limitations exist due to restrictions surrounding tool manufacture.

A concave blend radius is required when transitioning between a convex optical surface and planar surface such as a mounting flange, and grinding and polishing equipment require a fairly large radius to form the transition due to the diameter of the polishing head. However, the required blend radius for transitioning between a concave optical surface and planar surface can be much smaller since the blend radius becomes convex.

The opposite situation exists with molding tools since the mold tool contains an inverse of the finished geometry. The tooling blend radius for molding convex and planar surfaces is also convex, and therefore presents few limitations other than to not inhibit glass flow. The greater challenge comes when molding to a specific SAG height on a concave surface, since the optical surface of the tool will be convex, and therefore the minimum radius will be limited by the particular manufacturing method. However, the operation for creating an optical tooling surface is much different for oxide and chalcogenide mold dies (Fig. 8.3).

When given the option, it is best to allow the molded surface to have some degree of power rather than designing with planar surfaces. This is because when molding planar surfaces, it is very difficult to ensure that glass flow is *always* from the center outward. As the softened glass encounters the planar tool, it often happens that the glass rolls over the surface rather than flowing across it. Planar surfaces can always be produced to higher accuracy and quality through traditional grind and polish methods. Therefore, allow some curvature to the tool surfaces for better molding results (Fig. 8.4).

Chamfers are not easily molded, nor can they be easily created on the molding tool. Therefore, an equivalent radius is preferred in place of chamfers for molded optics. Also, PLM processing that does not require post processing such as centering is the

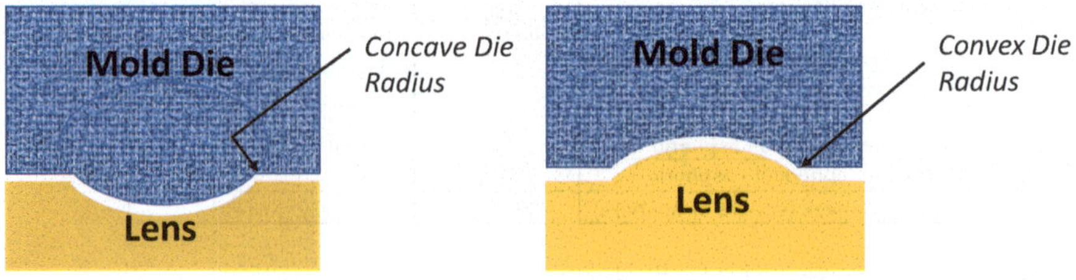

Fig. 8.3 Blend radii on mold dies for concave surfaces (l) and convex surfaces (r)

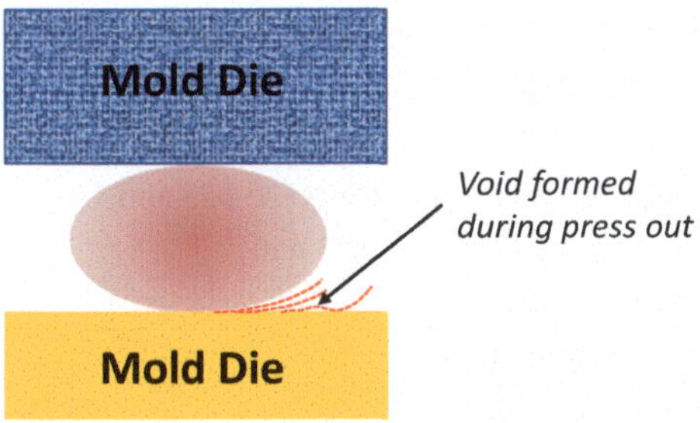

Fig. 8.4 Incomplete press out on planar surface

simplest and least expensive manufacturing option. Not only is the preform volume not held as tightly and is therefore less expensive, tooling cost is lower as well. When the diameter does not need to be tightly held and a small radius is allowable at its edge, consider using PLM processing that allows the diameter to free form.

8.2.2 Oxide Glass

When working with oxide glasses or high molding temperature glasses that require carbide tooling, a 2.50 mm minimum blend radius is suggested for mold surfaces that transition from convex to planar (concave lens transition). While grinding tools smaller than 5.0 mm diameter are available, the grinding tool shaft becomes so small that it is prone to deflection under grinding loads and may produce chatter and unsatisfactory surface attributes. As the small shaft deflects under load, surface profile becomes more difficult to maintain. Also, the reduced surface area of the tool results in faster tool wear, and the surface must be monitored more frequently than when working with larger diameter tools.

While attributes of a particular glass type may yield slightly different results, a partial list of preferred guidelines for many moldable glasses may contain the following. Tighter tolerances may be achievable, but are dependent on glass type and specific geometry.

- PLM process, free form diameter; 1.0–1.5 mm radius on the diameter
- Blend radius from optical surface to flange; 2.5 mm minimum
- Surface figure; 1 λ
- Surface quality; 40–20
- Surface roughness; 2 nm.

Table 8.1 Common tolerances for molded optics

	Oxide glass	Chalcogenide glass
Diameter	5–30 mm	3–30 mm
Aspheric figure error (fringes @ 633 nm)	2	2
Irregularity (fringes @ 633 nm)	1	1
Vertex radius	±0.75%	±0.75%
Decenter (mm)	±0.010	±0.010
Wedge (arcmin)	3	3
Center thickness tolerance (mm)	±0.020	±0.020
Diameter tolerance (mm)	±0.020	±0.020
Surface quality	40–20	60–40

8.2.3 Chalcogenide Glass

Chalcogenide glass and glasses having low molding temperatures do not usually require carbide substrates, which allows additional options for manufacture of the mold dies (see Sect. 2.4). Sharper blend radii are possible on these tools since the mold die surfaces are often diamond turned; but blend radii of 0.50 mm minimum are still suggested to promote good flow of glass during the pressing cycle. As with oxide glasses, attributes of a particular chalcogenide glass may produce slightly different results. A partial list of preferred guidelines for many chalcogenide glasses may contain the following; tighter tolerances may be achievable, but are dependent on glass type and specific geometry (Table 8.1).

- PLM process, free form diameter; 1.0–1.5 mm radius
- Blend radius from optical surface to flange; 0.5 mm minimum
- Surface figure; 1 λ
- Surface quality; 60–40
- Surface roughness; 2 nm.

8.3 Planning the Process

A suggested outline of process planning steps are offered as customer requests are considered for manufacture and production processes are being planned. This is not intended to capture all aspects and special circumstances for every type of request; rather, it is meant to provide a guide to help navigate the many decisions that are made when moving from customer request to supplier quote to developing an effective and reliable production process (Table 8.2; Fig. 8.5).

- Examine the request (preferably provided in the manner of a formal drawing) for material type, lens shape and size, surface specifications, special features, and requested quantities. Be careful of requests having incomplete information. This will almost always cause difficulty and expense at some point in the process.

- Glass type determines both the tooling material and coating type that should be used (see Chaps. 2 and 4). If the glass is not a type that is commonly used, look into the short and long term availability of glass types, melt frequencies, or any optical analogs that may be available from another supplier (see Chap. 5). Be especially careful with requests that call for a specific glass type with "no substitutes", as this may create difficulty with availability.
- Lens shape defines tool shape and preform shape. Determine the preform shape needed to mold the final surfaces. Examine the amount of glass flow for both upper and lower tooling to ensure the preform allows for balanced flow during the press out cycle. Be careful of assuming glass flow will be constant and even across the tool (see Sect. 7.4). Establishing the final molding temperature to be at the high or low edge of the spectrum may be useful in creating the desired flow pattern. Adjusting the amount of soak time needed for the process is also an option.
- Lens size determines the tooling cavitation and system throughput, along with estimated cycle times. For those just starting in PLM technology, stay with those diameters and aspect ratios that fall within your comfort range. There are many nuances to molding very large or very small diameter lenses, as well as high aspect ratio articles. Stay within your comfort range—perform development separate from production, and don't allow customer demands to drive development.
- Surface type is a factor often overlooked that may have a major influence on whether or not to accept the order. Even if they can be molded, bi-aspheric lenses are very difficult to measure and verify that specifications such as wedge and decenter have been satisfied. Metrology is the most important aspect of both development and customer satisfaction and should be given high priority when considering customer requests.

Table 8.2 Preform shapes and general applications

Lens Geometry	Preform Type			
	Ball Preform	Plano Plano	Lenslet (Plano - Convex)	Lenslet (Bi-Convex)
Bi-Convex	•			•
Equal Meniscus	•		•	
Positive Meniscus	•		•	
Negative Meniscus	•		•	
Bi-Concave		•		

Surface Description	Metrology Risk	Comments
Positive or Negative - Both Surfaces Aspheric		Very high risk, special fixturing and process required
Meniscus - both aspheric	High	Very high risk, special fixturing and process required
Meniscus - spherical / aspheric		Moderate risk, special fixturing and process may be required
Positive or Negative - Both Surfaces Spherical		Standard interferometer
Meniscus - both spherical		Standard interferometer
Planar / Concave - Aspheric		CGH, aspheric interferometer, or profilometric line trace
Planar / Convex - Aspheric		CGH, aspheric interferometer, or profilometric line trace
Planar / Concave - Spherical	Low	Standard interferometer
Planar / Convex - Spherical		Standard interferometer
Planar Window		Standard interferometer

Surface Description	Metrology Cost	Comments
Positive or Negative - Both Surfaces Aspheric		Very high cost, special fixturing and process required
Meniscus - both aspheric	High	Very high cost, special fixturing and process required
Meniscus - spherical / aspheric		Moderate cost, special fixturing and process may be
Planar / Concave - Aspheric		CGH, aspheric interferometer, or profilometric line trace
Planar / Convex - Aspheric		CGH, aspheric interferometer, or profilometric line trace
Positive or Negative - Both Surfaces Spherical		Standard interferometer
Meniscus - both spherical		Standard interferometer
Planar / Concave - Spherical	Low	Standard interferometer
Planar / Convex - Spherical		Standard interferometer
Planar Window		Standard interferometer

Fig. 8.5 Metrology guide for PLM operations

- Surface specs determine the quality of tool surface, quality of preform surface, and type of preform to use, and the decision whether to use a more expensive near net shape preform or a less expensive sphere, gob, or planar shape. When considering surface quality demands, the particular glass type often impacts the ability to meet specifications. Glasses having higher hardness are less prone to scratches and external damage. Always question whether the customer request can be achieved with the level of experience and molding technology available to the supplier.
- Special features such as diffractives or non-rotationally symmetric surfaces should be carefully considered. Much information exists to guide development efforts for creating special features, but information can never be a substitute for experience, so stay within your knowledge base and capabilities.
- Immediate, annual, and lifetime quantities are a factor when determining tool material and fabrication methods. For producing customer samples or prototype quantities, a single cavity mold can often be used to meet demands. Not only does it represent the simplest and least expensive type of tooling package with the lowest upfront costs and risks, but it also results in the lowest possible process times due to its low thermal mass. Once single cavity systems no longer meet quantity demands, multiple cavity tooling should be explored. Not all cavities need to be populated with finished tooling at the start. Ramp up tooling needs and related expenses as customer demand increases since the promise of future quantities is never certain.

• Calculate preform volume. Determine if the glass type is available in the required geometry (sphere, gob, window, near net shape, etc.). Estimate preform costs. The simplest way to obtain the actual volume of a lens is to analyze it in a software program such as Zemax, but when such tools are not available, one method of calculation is to break the lens into parts and derive a volume for each segment. An example of this is shown in Table 8.3 for a lens with convex surface 1 and concave surface 2.

• Estimate cycle times based on glass type, tooling materials, heat source, and thermal mass (Chaps. 2 and 3). Determine cavitation requirements based on requested quantities, cycle times, yields, number of shifts, etc. Estimate tooling costs and

Table 8.3 Example spreadsheet for calculating preform volume

Enter aperture surface 1	6.000	Enter aperture surface 2	6.000
Enter lens diameter	6.250	Enter lens center thickness	1.750
Enter <u>cx</u> aspheric (vertex) radius	5.000	Enter <u>cc</u> spherical (vertex) radius 2	28.50000
Enter conic constant k	0	Enter conic constant k	0
Enter D coefficient (4th power)	0.00000E+00	Enter D coefficient (4th power)	0.00E+00
Enter E coefficient (6th power)	0.00000E+00	Enter E coefficient (6th power)	0.00E+00
Enter F coefficient (8th power)	0.00000E+00	Enter F coefficient (8th power)	0.00E+00
Enter G coefficient (10th power)	0.00000E+00	Enter G coefficient (10th power)	0.00E+00
Enter H coefficient (12th power)	0.00000E+00	Enter H coefficient (12th power)	0.00E+00
Enter J coefficient (14th power)	0.00000E+00	Enter J coefficient (14th power)	0.00E+00
Enter K coefficient (16th power)	0.00000E+00	Enter K coefficient (16th power)	0.00E+00
Enter L coefficient (18th power)	0.00000E+00	Enter L coefficient (18th power)	0.00E+00
Enter M coefficient (20th power)	0.00000E+00	Enter M coefficient (20th power)	0.00E+00
Enter chamfer LW for side 1	0.150	Enter chamfer LW for side 2	0.015
Calculate "C" value (1/radius)	0.2	Calculate "C" value (1/radius)	0.035087719

$SAG_{cx\ surface1}$ =	1.000000			
$V_{cx\ surface1}$ =	14.661			
$SAG_{cc\ surface2}$ =	0.158335			
$V_{cc\ surface2}$ =	2.240			
$EDGE\ THK_{disc}$ =	2.592			
V_{disc} =	79.511			
Chamfer Vol. =	0.436			
			mm	gms
V_{total} =	66.655	Preform Ball Diameter "D" =	5.031	0.328

refurbishment costs (Sect. 2.4). Estimate throughput, number of parts per day, number of shifts needed, number of machines, robot feeders, aids to high cavity tooling, etc. (Table 8.4).

- Available equipment. Determine if existing equipment can produce parts to meet customer requirements, or whether modifications are needed. Can a high aspect ratio window or lens be molded on an IR or RF machine? Calculate (or estimate) potential thermal gradients and determine their impact on the specific glass type (Sect. 3.2.2). Do volumes require automated cleaning systems, or is manual cleaning acceptable? What types of chemicals are needed for cleaning, and are they based on glass type (Sect. 5.5)? Will special handling be required due to chemical reactivity of the particular glass type (Sect. 5.3.3)?

Table 8.4 Estimated tooling cost for various tooling materials and order quantities

	Tooling cost per cavity. Assumes 1-2 cavity mold.				Overall Tool Costs for xx Units				
Material	Mold Tool Cost ($)	Refurb Cost ($)	Cycles Bet Refurb	Possible Refurb Cycles	10,000	25,000	50,000	75,000	100,000
WC or SiC	$7,000	$3,000	5000	5	$13,000	$22,000	$41,000	$60,000	$79,000
Composite	$4,000	$1,500	3000	3	$8,500	$21,000	$38,000	$56,500	$73,500
Low Cost	$2,500	$1,000	3000	5	$5,500	$12,000	$21,500	$33,500	$43,000
Other	$2,500	$1,000	3000	5	$5,500	$12,000	$21,500		

** Based on 1-2 cavity tooling, assumes identical support structures in place*

- Look for unique features. Determine if FLM or PLM is best for your application (Sects. 2.2.3 and 2.2.4). What post processes are needed? Centering? Segmenting? Cut SAG? Cut flats?
- Design the tool package and mold surfaces (Sect. 3.6). Are special features required in the tool package (Sect. 2.2), should an insert sleeve be designed into the system to mold multiple sized lenses from a single mold die, are heat concentrators needed to minimize thermal gradients, etc.? Does the mold design contain shims, or will position control be used to manage CT tolerances (Sect. 2.2.2)? A sample spreadsheet may contain some of the information shown in Table 8.5.
- Specify coatings for tools (Chap. 4) and preforms (Sect. 4.4). Investigate stoichiometry of glass type to determine risk of interactions between materials.
- Define process parameters and design the molding process (Chap. 7). Be aware of any potential for devitrification (Chap. 6). Should vacuum molding be considered (Sect. 7.4)? Is the processing window of the glass short or long (see Sect. 3.2.3)? Every glass will mold over a range of temperatures and combined pressing force. Understand the impacts of using higher molding temperatures with lower force, or lower temperatures with higher force.
- When all aspects are considered and information is gathered, compile data and enter into the quoting tool. A sample quoting tool is shown in Table 8.6, and may contain some of the features shown.
- Go back and check the request for additional information that may have been overlooked. For example, transmission of the finished lens after coating may not be possible due to absorption in the material. This is especially true when working in the infrared, since requests may sometimes cover a wide range of wavelengths, and the transmission of chalcogenide glasses can change greatly at the edges of their transmission bands (ex., 8–13 μm wavelength requests with $As_{40}Se_{60}$ chalcogenide glass, where T drops rapidly after 12 μm) (Fig. 8.6).

8.4 Conclusion

Precision molding equipment, technologies that support glass molding, and process development efforts have been driving the advancement of molded glass optics for over a century. Without question, glass molding could not have advanced to its current state without innovations in machining, coating, metrology, and material science. The continued development of materials and processing techniques promises to deliver exciting challenges and opportunities for years to come.

Glass molding has found application with large and small quantities, large and small diameters, and a variety of surface geometries to offer a low cost alternative for the accurate and reliable production of optical surfaces. But there also exists a need for good process understanding and consideration of non-conventional methods of production. It is important to learn from every situation and apply that knowledge to future opportunities.

Table 8.5 Sample tool package design calculations

SURFACE 2	Lens Surface 2	Tool Surface 2
Enter Clear Aperture	23.800	23.858039
Enter "C" Coefficient (1/Radius)	0.011039	0.011012
Enter "k" Value	0.000000	0.000000
Enter "D" Coefficient (4th Power)	-2.48246E-06	-2.46439E-06
Enter "E" Coefficient (6th Power)	-1.76972E-09	-1.74830E-09
Enter "F" Coefficient (8th Power)	-1.75209E-12	-1.72247E-12
Enter "G" Coefficient (10th Power)	0.00000E+00	0.00000E+00
Enter "H" Coefficient (12th Power)	0.00000E+00	0.00000E+00
Enter "J" Coefficient (14th Power)	0.00000E+00	0.00000E+00
Enter "K" Coefficient (16th Power)	0.00000E+00	0.00000E+00
Enter "L" Coefficient (18th Power)	0.00000E+00	0.00000E+00
Enter "M" Coefficient (20th Power)	0.00000E+00	0.00000E+00
Enter Preform Radius	6.037	
Lens Base Radius (Calculated)	90.5900	
Tool Base Radius (Calculated)		90.81091

Diffractive Surfaces	Lens Surface 2	Tool Surface 2
Enter design wavelength (mm)	1.00000E-02	
Enter index @ λ_o (mm)	2.76608E+00	
Enter 1st phase coefficient (ρ^2)	-3.15636E-04	-3.14868E-04
Enter 2nd phase coefficient (ρ^4)		0.00000E+00
Enter 3rd phase coefficient (ρ^6)		0.00000E+00
Enter 4th phase coefficient (ρ^8)		0.00000E+00
Enter 5th phase coefficient (ρ^{10})		0.00000E+00
Calculate Step Height (mm)	5.66226E-03	5.67607E-03
Diffractive Zones		
Enter zone radius for ring 1	5.629	5.6427
Enter zone radius for ring 2	7.960	7.9794
Enter zone radius for ring 3	9.749	9.7728
Enter zone radius for ring 4	11.257	11.2845
Enter zone radius for ring 5		0.0000

Step 1: Core diameter	
Enter Desired Lens Diameter	3.5
ChG Lens Diameter at Tp	3.515
Lower Core Diameter at Room	3.504
Clearance for Upper Tool	0.004
Upper Core Diameter at Room	3.500

Step 2: Sleeve bore diameter	
Lower Tool Clearance at Tp	0.000
Sleeve Bore Diameter at Room	3.509

Step 3: Outer sleeve diameter	
Sleeve Clearance at Tp	0.001
Enter Max Die Diameter	36.001
Die Diameter at Tp	36.053
Outer Sleeve Diameter	35.984

Step 4: Tool Heights	
Enter Max Upper Die Height	15.000
Enter Max Lower Die Height	18.000
Max Upper Die at Tp	15.023
Max Lower Die at Tp	18.026
Enter CT of the lens	1.1
Enter Total Height of the lens	1.1
Enter Desired Parting Drop	0.1
Lower Core CT at Tp	16.821
Enter Parting Line Spacing	0.5
Upper Core CT at Tp	15.628
Calculated Core Heigl	33.549
Calculated Tool Height at Tp	33.549
Lower Core CT at Room	16.761
Upper Core CT at Room	15.575

Table 8.6 Sample spreadsheet useful as a quoting tool

Estimated Number of Cavities	4
Estimated PLM Cycle Time (min)	25

Chemical Properties	
Water Resistance(Powder) Group RW(P)	1
Acid Resistance(Powder) Group RA(P)	3
Weathering Resistance(Surface) Group W(S)	1
Acid Resistance(Surface) Group SR	4.2
Phosphate Resistance PR	1.0

Thermal Properties		
Strain Point StP (°C)		
Annealing Point AP (°C)		
Transformation Temperature Tg (°C)		612
Yield Point At (°C)		632
Softening Point SP (°C)		676
Expansion Coefficients	(-30~+70°C)	56
α (10⁻⁷/°C)	(+100~+300°C)	71
Thermal Conductivity k W/(m·K)		0.872

Operation	Process Steps	Set Up Min	Std Min EA	Labor Cost	Material Cost
Purchase Preform	Purchasing			$0.00	$5.50
Incoming Inspection	Verify qty, physical dims, mat'l specs & CoC	0.20	1.00	$2.00	
Preform Cleaning	U/S cleaning		0.30	$0.50	
Preform Handling	Set up trays, load material	0.20	0.25	$0.75	
Mold Tool Set Up	Set up tooling	0.09		$0.15	
PLM Processing	Machine time		6.38	$10.63	
PLM Metrology	Physical dimensions, visual inspection, SQ, each lens		1.00	$1.67	
PLM Metrology	Surface figure & roughness, 1 lens per every 4		3.75	$6.25	
Packaging	Package, ship to centering dept.		0.50	$0.83	
Centering Set Up	Set up tooling	0.18		$0.30	
Centering	Machine Time		5.52	$9.20	
Centering Metrology	Physical dimensions, visual inspection, SQ, each lens		1.00	$1.67	
Packaging	Prepare for transfer to final inspection		0.10	$0.17	
Inspection	Document qty, physical dims, mat'l specs & CoC	0.30	0.30	$1.00	
Final Metrology	Compile data, spreadsheet, burn CD		0.30	$0.50	
Packaging	Package, ship		0.10	$0.17	
				PLM	*FLM*
		Labor Cross Check =		$35.78	$22.78

Expected Yield	98%
Order Quantity	25000
Total Including Loss	25,511
Lot Size	100
NRE FLM Tool cost	$32,000.00
NRE Robot Tool Cost	$0.00
Final diameter (mm)	23.000
Final height (mm)	5.130
Blank Volume (cm³)	6.86E-02
Total matl req'd (cm³)	1.75E+03
MATERIAL TYPE	
Material Cost ($/cm³)	
Mat'l (Blank $/Order)	$ -
Mat'l (Blank $/Pc)	$ -

Comments / Notes

Assumes labor rate = $100/hour. Assumes 35.0mm OD single cavity tooling; 4 cavity assumes 42.0mm B and 80.0mm OD tooling. Assumes 100% inspection for single cavity and for lot sizes of 50 pieces or less, AQL inspection for lot sizes above 50 pieces. Assumes U/S Cleaning lot size of 100 pieces maximum. Assumes 100 pieces per tray. Assumes use of ground & polished ball preforms. Assumes max tool life of cycles between tear down, cleaning, & inspection. Assumes $3,500 per tool plus $500 for sleeve, $2,000 each for 4 cavity tooling. Assumes one (1) complete spare set of tools for both single and 4 cavity tool se

Total Set Up Min EACH	0.97				
Total Run Min EACH	20.50	Labor Rate			
Total STD MIN EACH	21.47	$ 1.67	$	35.78	
Total Estimated UMC EACH			$	41.28	$ 28.28
Total Estimated UMC EACH w/NRE Amortized			$	48.06	$ 29.56

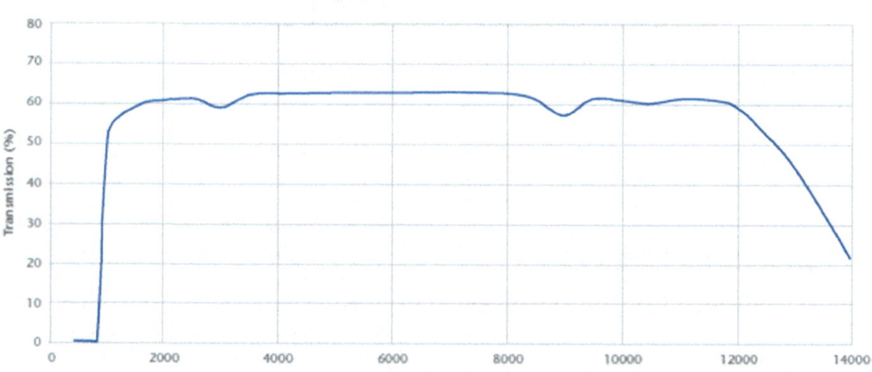

Fig. 8.6 Transmission of $As_{40}Se_{60}$ chalcogenide glass [1]

Good data is an absolute necessity in order to produce good parts, but accurate information regarding material properties may not be easy to find. Room temperature values can be found for many physical and thermal properties, but accurate data at higher temps may be more difficult to obtain (CTE values between room temperature and process temperatures, thermal conductivity at critical temperatures, etc.). When this information is not available, some learnings can be gained empirically that will produce better processes in the future.

Disappointments during development are not a bad thing—they help guide us to future learnings and discoveries, and get us focused in the right direction. Failure is the place where learning begins, and should be embraced as evidence that we are pushing the boundaries of our current knowledge base and are about to learn something new. Education is the foundation for learning, but education is not an end in itself. Whether through a textbook, a white paper, or in a laboratory, if we stop learning when the course is finished, then we've learned nothing at all. Process development is a means of continual learning, and it is our hope that the information from these pages will be helpful for meeting your special molding requirements and become a valued source of material for exploring new opportunities in glass molding.

Reference

1. SCHOTT infrared chalcogenide glasses irg, 26 October 2013, US SCHOTT AG, 2013

The manufacturer's authorised representative in the EU is Springer Nature Customer Service Centre GmbH, Europaplatz 3, 69115 Heidelberg, Germany. If you have any concerns regarding our products, please contact ProductSafety@springernature.com

Printed and bound by CPI Group (UK) Ltd, Croydon, CR0 4YY

30/09/2026

02225574-0012